J M Leonard M A
Head of the Mathematics Department
King Edward VI School, Lichfield

Statistics
The arithmetic of decision-making

The English Universities Press Ltd

ISBN 0 340 12485 7

First published 1971

The English Universities Press Ltd
St Paul's House Warwick Lane London EC4P 4AH

Filmset by Keyspools Ltd, Golborne, Lancs
Printed and bound in Great Britain by C. Tinling & Co. Ltd,
London and Prescot

Preface

As the State becomes more highly organised and more interested in the scientific analysis of its life, there appears an urgent necessity for various statistical information, and this can be properly obtained, reduced, correlated and interpreted only when the guiding spirits in the work have the necessary mathematical training in the theory of statistics. Figures may not lie, but statistics compiled unscientifically and analysed incompetently are almost sure to be misleading, and when this condition is unnecessarily chronic the so-called statisticians may well be called liars.

E B Wilson (1912)

The aim of this book, to put it at its lowest, is to cover the Advanced Level syllabuses in Statistics of the major Examining Boards. These syllabuses are not, of course, all the same, nor is it likely that they will all remain in their present form for very long, so that omissions (mostly of a minor nature) will be found. But, for some time to come, it should provide the majority of the material necessary, either for the student working on his own or for a teacher with a group.

This was the reason that the book came to be written—by a practising sixth form teacher who found the existing published material inadequate and thought that he might be able to do better. But the needs of other possible readers have been kept in mind also: in particular, university and college students, who may be studying such diverse subjects as medicine, geography, economics, social science, engineering or government administration, and who find that statistics forms part of their course. Often these readers will be able to skip the mathematical sections without loss of continuity and, at appropriate points, this is indicated in the text.

But let those who read the book also raise their sights a little, above such trivialities as examinations, and try to see something of statistics on its merits—merits which are far more considerable than is often thought. It must be admitted that the subject does not get a good press, and that there are good reasons why this should be so. But those who survive at least the first eleven chapters should be in a good position to defend the trustworthiness of the subject, and to expose the fallacies

iii

03365

which so often appear in its name. If the theme of the book can be expressed in a single sentence, it is this—that statistics deals with the mathematics of the real world.

I am grateful to Messrs D. V. Lindley and J. C. P. Miller, and to the Cambridge University Press, for permission to reproduce tables 2 and 4 from the *Cambridge Elementary Statistical Tables,* and to Mr J. K. Backhouse and Messrs Longmans, Green and Co. Ltd, for permission to reproduce table A from *Statistics: an Introduction to Tests of Significance.* I am also indebted to the Literary Executor of the late Sir Ronald A. Fisher, to Dr Frank Yates, F.R.S., and to Oliver and Boyd, Ltd, Edinburgh, for permission to reprint tables III and IV from their book *Statistical Tables for Biological, Agricultural and Medical Research.* The majority of the other tables are taken by permission from the *Lanchester Short Statistical Tables,* by G. R. Braithwaite and C. O. D. Titmus, published by the English Universities Press.

The various Examining Boards have all been most helpful in giving permission to reprint questions from their papers. I am grateful to the Associated Examining Board, the Joint Matriculation Board, the Oxford and Cambridge Schools Examination Board, and the Oxford Delegacy of Local Examinations. Individual questions from these sources are indicated by appropriate abbreviations.

Many other people have helped in the making of the book. I am grateful to my former colleagues at Manchester Grammar School, particularly Philip Schofield, and to my pupils there on whom much of the material has been tried out; to Peter Holmes of Doncaster College of Education; to Mr A. Duval, Chief Actuary of the Co-operative Insurance Society, and to members of his staff, particularly Mr P. D. Johnson; to Professor M. B. Priestley of the University of Manchester Institute of Science and Technology; and, finally, to Roger Stone, for his unfailing advice and encouragement. Without each one of these, the book would have been in some measure the poorer; its remaining weaknesses (and any errors which have got past the attempts to weed them out) are mine.

J M L

Contents

Chapter 1

Introduction

The title of this book reflects the main purpose of the study of statistics—that it is fundamental to good decision-making. The decisions may be major ones on which the lives or the welfare of many people depend, or they may be quite trivial. An example will illustrate the point.

At the time of writing, there is much debate over the question of comprehensive schools: whether to 'go comprehensive' and if so in what form. One of the chief discussion points is whether educational standards in the sixth form are different under a comprehensive system, as a result of having less selection. If this were settled positively one way or the other, the whole matter could be seen in a much clearer light. Such a comparison might seem straightforward enough—one works out the two sets of figures, and if there is a reasonable margin between them, there is the answer. But in fact it is not as easy as that.

First, we must examine the data with some care. We cannot possibly take into account every single boy and girl who has been through a sixth form in, say, the last five years—there are too many of them. So we must take a representative selection, or sample, from each type of school, and in doing so we introduce two possible sources of error: the samples might be too small to give a balanced picture, or they might be unrepresentative. Then we must ensure that the measures used for comparison are in fact the right ones—A-levels are no doubt important, but they are not the only measure of educational standards. These are some, but not all, of the pitfalls.

Further, we must be rather careful about what constitutes a reasonable margin of difference, and how it should be interpreted. Partly because of the errors involved in sampling, and partly because of the nature of the figures themselves, there is bound to be at least a small difference, and it is not always obvious when it becomes important.

This problem is typical of the work of a statistician. Before starting any calculations, he must ensure that the two sets of figures are comparable; he can then apply a suitable test and state his conclusions. Common sense is not usually enough—only very rarely will an answer be directly contrary to what common sense would predict, and it is often temptingly easy to guess what an answer should be, but it is not so easy to guess anywhere near right.

As one consequence of this, the man in the street often holds the subject in low regard, because he thinks that his own intuitive answer is better than that of the 'expert'. He is often misled also by an all-too-well-justified suspicion that the figures are being used to prove a point in a one-sided manner, when a disinterested observer might well reach a quite different conclusion.

In this book, the intellectual respectability of the study of statistics will be vigorously defended—the statistician is as well qualified as the pure mathematician to state his conclusions with confidence and precision and with their limitations clearly defined. Needless to say, this ideal is not always attained, and there can be few subjects in which a little learning is such a dangerous thing, or is so often applied. The misuse of statistics in advertising in recent years has contributed further to the low esteem in which the subject is held by the man in the street, but this must not preclude the right use of it.

The word 'statistics' is derived from 'state', since it was originally concerned with national figures for population and wealth. (This got it off to a bad start in the popularity stakes, because the reason for obtaining the figures was to raise more taxes.) The word is now of course used much more widely, and in three different senses:
(a) as a plural noun, statistics are large numbers of figures, representing measurements of physical quantities;
(b) as a collective noun, statistics is the study of such figures;
(c) as a singular noun, 'statistic' is used in a specialised way which will be defined later on.
If this seems confusing, do not worry—the context always makes the meaning clear. The commonest use in this book will be the second one, and a formal definition of it may be given as follows:

Statistics is the science of collecting and analysing numerical data, with the aim of making optimum use of the information contained in the figures in reaching decisions about a future course of action.

The course of action involved may be one's own or somebody else's: statistics in advertising are intended to make the consumer buy the product advertised. It is, of course, possible, and all too common, to present one's material in a biased manner, whether consciously or unconsciously, and all statistical data should be treated with suspicion until clearly shown to be reliable.

Roughly, the way in which the subject is developed in this book is as follows. The theory of probability gives the mathematical basis for the processes of decision-making, and this is introduced in chapter 4. The remainder of the book deals with these processes in detail—the various 'tests of significance' and the circumstances in which they are used. The first three chapters lay the necessary groundwork, under the general title of Descriptive Statistics. This chapter itself covers the simple but essential preliminaries, the methods of summarising statistical data by grouping figures together and presenting the results in tabular or pictorial form.

Definitions

One point of difficulty at this stage is the fact that statisticians use a number of familiar words in a rather specialised sense, as well as other technical terms of their own. You will find that it takes a little time to get used to these terms, and some mental effort as well. We start by defining a few of them.

Population: a given set of objects under consideration, not necessarily people.

Sample: a number of objects selected from a population.

Variable: a quantity which may be counted or measured. Variables may be discrete (measured in indivisible units or steps, like numbers of eggs) or continuous (like time). In practice all variables may be taken as discrete because continuous variables can only be measured to finite accuracy, but the concept of a continuous variable is useful in mathematical analysis. Variables need not be numerical: categories such as colours, sexes or suits in cards may be used as well.

Variate: the term used for a typical measurement of a variable. The distinction between a variable and a variate is a fine one, often ignored in practice. But it is useful nevertheless, and worth preserving.

Distribution: for a given set of variates, a term which implies that the relative numbers of each possible measurement are being considered. (This is a rough definition only—we shall return to it later.)

You will probably need to refer back to these definitions as the terms occur. To help make the meanings clear, suppose we were investigating the weights of a given batch of potatoes. The *population* is the set of potatoes, in no particular order. The *variable* is the weight, and the *variate* is a typical figure for the weight of one potato. The variate has a certain *distribution*—that is, we could show on a graph the relative numbers of potatoes within each chosen range of weights. This could be done either for the whole population or for a *sample* drawn from the population.

Tabulation of data

The raw data—the first figures which go down on paper in any statistical investigation—are usually lengthy, and must be summarised in some way before making use of them. The first step in this is to extract the information which is likely to be useful and ignore the rest. The reader who is likely to be concerned only with the type of statistical problem set in examinations will not, unfortunately, have to do this very much. But in practical statistics it is important, because few organisations keep their records in the most suitable form for subsequent statistical analysis.

After any necessary selection has been done, the figures are grouped together. This is mainly a matter of common sense, but a few hints may be useful.

(a) The number of groups or **classes** will depend on the purpose and thus the accuracy of the investigation. It will rarely be fewer than six, and not often more than twenty. If in doubt, use the smaller class size, and combine adjacent ones together later if it proves possible to do so.

(b) The accuracy of the original data will also affect the choice of class size. The more precise the data, the smaller the best class size will be.

(c) If the range of values covered by each group (the **class interval**) is 10, it is better to take say 40 to 49 than 41 to 50. In counting up, the last digits can then be ignored.

(d) Counting up is best done by writing for each figure in the data a mark ı in the appropriate column, and for each fifth item in the same category a horizontal line through the last four, as ⊩. Mistakes are easily made at this stage, and also difficult to trace, so that a check after each hundred figures of the data is recommended.

(e) It is essential to know, and to write down, the way in which the original measurements were made. This is because the half-way point in the class, which is usually taken to be the average value of the variates in that class, depends on the method of measurement.

For example, if ages are measured in completed years, the average of the 40 to 44 group is taken as 42·5, because everyone *between* ages 40 and 45 is in this group. But if ages are measured to the nearest year, the average of the 40 to 44 group will be 42 exactly.

(f) When data has been grouped, it is necessary to remember that information has been lost in the process. With experience, you will be able to judge the amount of inaccuracy in a calculation using grouped data, so that answers may be given to a suitable number of significant figures.

The histogram

The next stage after grouping the data is to present it in a pictorial form called a histogram. Even if a detailed calculation is to follow, a histogram will often help both in visualising the problem and in avoiding answers which are wildly wrong. The horizontal axis is marked in the units of the data, showing the class divisions. The number of observations in each class (the **frequency**) is then shown by a block whose *area* is proportional to that number. When a uniform class interval is used, the *height* of the block is also proportional to the frequency, and the vertical axis can then be marked in suitable units. But sometimes it is convenient to use unequal class intervals: for example, small intervals in the centre of the distribution where there are many observations, and larger intervals at the ends where there are fewer. Also, the concept of *area* representing *frequency* will be important later on.

The examples in figures 1.1 and 1.2 explain themselves, but some small points may be noted.

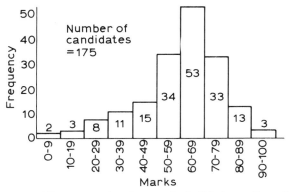

Figure 1.1   Histogram showing marks in third year French exam

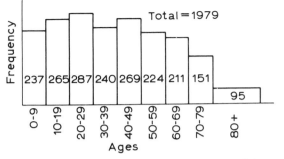

Figure 1.2   Histogram showing the age distribution of the population
of Middletown on April 25th 1971

(a) Titles are given, and axes marked where appropriate.
(b) Frequencies are shown by numbers as well as graphically: this is not always necessary but may aid interpretation.
(c) Note the unequal class intervals: in the last group of marks there are 11 marks in the group, which is hardly worth bothering about, and the last group of ages is not given a specific limit but set nominally at 100.
(d) The first distribution shows a tendency for the marks to be concentrated around the middle; this is a common feature of many distributions. The reason why the other is a quite different shape will be clear enough.
(e) The numbers of classes are ten and nine; these are typical figures, and the number will almost always be in the range six to twenty.

Figure 1.3 is slightly different. Instead of showing the numbers of men in each range of heights, it shows the proportion of the total in each case:

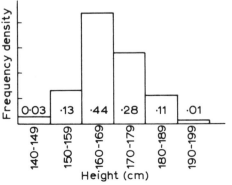

Figure 1.3   Frequency density histogram showing the heights to the
nearest centimetre of 631 adult men

for example, 21 men out of 631 is shown as 0·03 to two decimal places. The shape of the histogram is unaffected. The total area of the histogram, measured on the scale being used, must be unity, and this enables two histograms covering different data to be compared. For example, the heights of 400 women (or 4000) could be shown on the same histogram in a distinctive colour, giving a visual comparison between the two distributions, irrespective of the different totals. A diagram of this type is called a **frequency density histogram**.

The frequency polygon

The mid-points of the top edges of each block of a histogram can be joined by a series of straight lines, and the result is called a frequency polygon. It is sometimes useful when comparing two distributions on the same diagram, but in general the histogram is to be preferred.

Frequency curves and frequency density curves

As a variation of the frequency polygon, the mid-points of the top edges may be joined by a smooth curve called a **frequency curve**. For a large population which gives a histogram of a regular shape, the curve will be a better representation of the actual distribution than a histogram, because the blocks of the histogram have been smoothed out. For a smaller population it may be necessary to smooth out some of the irregularities in the histogram as the curve is drawn, and the result will be an approximation which shows only the general shape of the distribution. But it is useful nevertheless, and has the simple merit of being very quick to draw.

The cumulative frequency curve or ogive

This is a graph of the cumulative number of observations (that is, the total number up to a particular point—compare the word accumulator) against the variable. Figure 1.4 shows the data of figure 1.1 plotted in this way.

The shape of the curve is a distinctive one, and the alternative name of ogive (pronounced with a soft 'g') is the technical name for an architectural feature of this shape. The points on the graph may be joined by straight lines, making a cumulative frequency polygon, but a curve is preferable. The main use of the ogive is in estimating medians and quantiles, which will be mentioned in the next chapter; it does not give as clear a picture of the shape of the distribution as a histogram or a frequency curve.

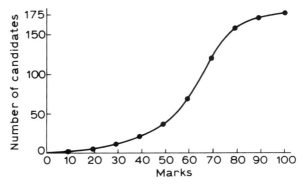

Figure 1.4   Cumulative frequency curve of marks in third year French exam

Two points to notice are that the frequency is always shown on the vertical axis, and that the point on the graph for, say, the number of candidates up to 59 marks (73) is above the figure 59, and not above the mid-point of the 50 to 59 block as it would be in a frequency polygon. Strictly speaking a curve is not justifiable in this instance, since a mark total is a discrete variable, but pedantic objections of this nature can safely be ignored.

Displaying data by categories

Many types of statistical data are best classified in categories, rather than by grouping variates in ranges of values. For example, government expenditure is divided into defence, education, health and so on. Such information is often shown pictorially, and three methods of doing this are given here: they are the **bar diagram**, the **pie chart** and the **pictogram**. Figures 1.5 and 1.6 represent the same data.

The bar diagram is similar in appearance to the histogram, although the bars are often shown with gaps between them. Its main use is to show the relationships of the various categories to each other rather than to the whole.

The pie chart is so called because it looks like a pie divided into sections. Here the immediate visual effect is to show the relationship of each part to the whole: if one category accounts for about a quarter of the total amount, this is clear at a glance. Comparing the relationship of one category to another is more difficult, particularly if they are on opposite sides of the pie. The number of categories must be fairly small to avoid congestion, and categories accounting for less than about two per cent of the total are difficult to show clearly.

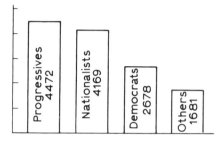

Figure 1.5   Bar diagram showing election result in Barsetshire North

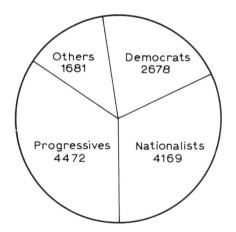

Figure 1.6   Pie chart showing election result in Barsetshire North

The pictogram is a variation of the bar chart, used for added visual attraction. It is most used in printed work where the symbols can be reproduced with ease; in manuscript the drawing process can be tedious. As a further variation, pictures of different sizes can be used, but care is necessary as the human eye naturally compares areas rather than linear dimensions. For example, a square of side two centimetres *looks* about four times the size of a square of side one centimetre—certainly far more than twice as big.

Twenty viewers tuned in

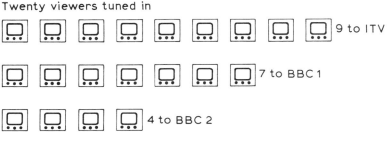

Figure 1.7   Pictogram showing the proportions of viewers tuned to three television channels at a given time

Time series diagrams

The commonest pictorial method of displaying statistical data is the time series diagram, which shows the different values of a variable over a period of time. Two examples are shown in figures 1.8 and 1.9, and again the diagrams more or less explain themselves. The values may be shown either by joining a series of points by straight lines, as in a frequency polygon, or by blocks for the separate years or months. Some quantities are measured as totals over a period of time (rainfall, for example), and

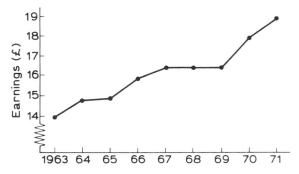

Figure 1.8   Time series showing mean weekly earnings of employees of Bloggs Plastic Toy Company

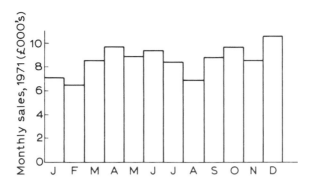

Figure 1.9   Time series showing monthly sales for 1971 in thousands of pounds for Bloggs Plastic Toy Company

here a block diagram is more appropriate; others can be assessed at any convenient moment in time (a cost-of-living index, for example), and here a polygon would be better.

Figure 1.8 has a false zero, and this can be shown either by a dotted section at the foot of the axis, or as here by a zig-zag section. Predicted values are sometimes shown dotted.

The purpose of a time series diagram is to give an immediate visual impression of the trend of the changes and also any cyclic variations there may be. For example, a diagram showing monthly figures for foreign exchange earnings from tourism over five years would show both: a trend upwards because tourism is generally increasing, combined with a seasonal variation because most tourists travel in summer. The trend could be shown more clearly by plotting annual totals only, so that seasonal variations do not show. But the two can be separated much better by means of a moving average, which will be mentioned in the next chapter.

The exponential curve

In order to round off the chapter, we shall take a brief look at a time series of rather special importance. When a time series is governed by unchanging factors so that it shows a regular trend, it is often not a straight line graph but a curve. Compound interest is an example, assuming that the rate of interest is constant. The rate of increase of the sum invested depends not only on the rate of interest, but also on the sum itself which is increasing all the time as interest accumulates. So the rate of increase also goes up: in other words, the graph gets steeper, even if the rate of interest is constant. The same is true in reverse for a decreasing quantity, for example radioactive decay. The shapes are shown in figure 1.10.

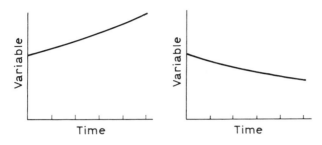

Figure 1.10   The exponential curve

Mathematically, the two are more or less the same: the decreasing one with its time scale reversed fits on to the end of the other one. The resulting curve is the exponential curve, important both in pure mathematics and in statistics. In mathematical terms, it is the graph of a variable whose rate of change is proportional to its own instantaneous value, and it will crop up in various guises many times in this book.

In a time series diagram of this kind, a logarithmic vertical scale is often used. This reduces an exponential curve to a straight line, so that variations from the exact exponential may be seen more clearly. The same *proportional* increase (e.g. 4 to 6 and 6 to 9 in two successive years) would then appear as a steady increase, but an increase from 9 to 13 in the next year would be shown, correctly, as a smaller rate of increase.

Exercises

1 Collect, and criticise, examples of pictorially displayed data from published journals.
2 The numbers of deals marked in certain ordinary shares on the London Stock Exchange on 50 business days were:

66	133	89	97	112	104	96	105	117	88
77	89	91	113	109	102	94	107	119	87
71	127	125	109	99	103	96	108	117	88
81	91	124	98	109	101	105	93	121	86
84	131	92	97	100	101	106	93	119	87

Draw up a frequency table showing the numbers in the intervals 60 to 69, 70 to 79, etc., and draw a histogram to illustrate the data. Draw also a cumulative frequency curve for the distribution.   (O & C)

3 For the data of question 2, consider and comment on the usefulness of a histogram in which the class interval is taken as (a) 1, (b) 2, (c) 5, (d) 20, and (e) 50, instead of 10. Is there anything to be said for taking unequal class intervals?

4 Draw histograms showing the total scores obtained by throwing two, three or four dice a large number of times.

5 Compare the literary styles of two authors by drawing histograms showing the distribution of numbers of words per sentence, in several hundred sentences from each. Alternatively, the numbers of letters per word may be used, or the number per 100 words of a common word such as 'and'. If done as a co-operative effort, the data may be used in a more objective way, to compare one author's variability within the same work and in different works with variability between different authors: in a highly developed form, this method can be used to investigate cases of disputed authorship.

Chapter 2

The average

Median, mode and mean

It is possible to say, without being misleading, that men are taller than women. In ordinary conversation this would be taken to mean not that all men are taller than all women, but that a typical man is likely to be taller than a typical woman. By 'typical', we mean about average height in the way the word is commonly understood. But there are several different kinds of average, and in addition there are other measures which serve a similar purpose. Later on in this book we shall be making statements of the type of 'men are taller than women' with very much greater precision of meaning, and the first step in the process is to define more clearly what is meant by an average in this context.

What we require is a single figure which is in some way representative of the distribution as a whole. One method is to arrange the variates in order of magnitude, as in the military system of 'tallest on the right, shortest on the left'. The one in the middle is then taken as representative, since it has as many variates larger than it as smaller. The typical variate defined in this way is the **median**. If there are 25 variates, the median is the 13th, with 12 each side of it; if there is an even number of variates, the median is taken to be half way between the values of the middle two. The general formula is that the median is the $\frac{1}{2}(n+1)$'th variate out of n.

The median is easy to calculate from a cumulative frequency curve: simply take the point on the horizontal axis corresponding to the point half way up the frequency axis. If the data is in the form of grouped figures or a histogram, the median can be estimated by simple proportion: using the data of figure 1.1 for example, the median mark is that of the 88th child in order, who is 15th (from the bottom) of the 53 in the 60–69 class. The variable has to be treated as continuous for calculation purposes, and so the class range is taken as $50\frac{1}{2}$ to $69\frac{1}{2}$. (This procedure is discussed

14

in more detail in chapter 4.) The median mark is then $59\frac{1}{2} + 10 \times 15/53$, or 62·3. Such accuracy is not usually needed, and often the change from 60 to $59\frac{1}{2}$ would be ignored—if we wanted to be really fussy, the 15 should be taken as $14\frac{1}{2}$ anyhow. If this is not considered accurate enough, a better measure than the median should be used instead.

The principal objection to the median is that it takes no account of asymmetry or irregularity in the distribution. To take an extreme case, a distribution of three 11's, three 12's and five 19's has a median of 12, which is far from representative. It is also difficult to use with a discrete variable which takes only a small number of values, for example children in a family. But in some fairly regular distributions it is useful, and easy to calculate.

For a continuous variable grouped into classes, the **median class** is sometimes used. As the name implies, it is the class which contains the median.

Another type of representative figure, the **mode**, is also simple to calculate but otherwise has little merit. For a discrete variable, it is the value of the most common variate, the one with the tallest block on the histogram. The obvious and overriding objection to the mode is that it is hopelessly misleading for irregular distributions. For a continuous variable, the mode is taken to be the peak of the density curve. The term **modal class** can also be used for a grouped distribution.

Much more commonly used than these, and for most purposes much better, is the arithmetical average, which will be referred to here as the **mean**. This is defined as the sum of all the variates divided by the number of variates, following the almost intuitive form that most people know— to find the average weight of a rowing crew, add the eight weights together and divide by eight. More refined methods of calculation will be discussed later.

The mean takes longer to calculate than the median or mode, but its value is affected by the size of every member of the distribution. In other words it is usually a more truly representative figure, particularly in asymmetrical or irregular distributions.

But there is one case where the median may give a truer picture than the mean—this is when a distribution has a few members of exceptional size. For example, if we are comparing salary levels for corresponding age groups in two professions, there might be a few individuals in one of the groups with exceptionally high salaries, who therefore made a misleadingly high contribution towards the mean. A person who considered himself 'about average' might be more interested in the median salary level.

Moving averages

Frequently a quantity measured at different points in time is subject to seasonal variations. Since the reason for making the measurements is normally to detect non-routine variations, some way of filtering out the effect of routine changes would be useful. The moving average is the most usual method.

If the quantity being measured is the quarterly output of new cars, the comparison between the fourth quarter and the third would not be very useful because car output is usually lower in summer owing to annual holidays. A better standard of comparison would be the fourth quarter of the previous year, and this is effectively what a moving average does. In this case a four-quarter moving average is the type used, because the figures are computed quarterly while the seasonal variations follow an annual cycle. For each quarter, the average of the four quarters ending with that one is calculated: an example will make the method clear.

	1st quarter	2nd quarter	3rd quarter	4th quarter
1966	196	227	184	241
1967	235	249	206	258
1968	216	234	171	232
1969	231	256	230	269
1970	240	277	228	273

Quarterly output of new cars in Ruritania (thousands)

	1st quarter	2nd quarter	3rd quarter	4th quarter
1966				212
1967	$221\frac{3}{4}$	$227\frac{1}{4}$	$232\frac{3}{4}$	237
1968	$232\frac{1}{4}$	$228\frac{1}{2}$	$219\frac{3}{4}$	$213\frac{1}{4}$
1969	217	$222\frac{1}{2}$	$237\frac{1}{4}$	$246\frac{1}{2}$
1970	$248\frac{1}{2}$	254	$253\frac{1}{2}$	$254\frac{1}{2}$

Four-quarter moving average for output of new cars in
Ruritania (thousands)

The data is shown in pictorial form in figure 2.1. The trend is visible on the graph of the actual quarterly figures if one looks hard enough, but it is very much clearer on the graph of the moving average.

The easiest method of calculation is as follows. First compute the mean of the first four quarters, which is 212. For the next figure, the output for the fifth quarter has to be added to the total and the first taken off, before dividing by four. The increase from the previous figure is therefore one-quarter of the difference between 196 and 235, which is $9\frac{3}{4}$, and the average is thus 221¾. In some cases, of course, there is a decrease. An independent check of the last four quarters is essential, and in a lengthy calculation further intermediate checks will save much frustration.

This method shows, incidentally, why the effect of a moving average is to indicate changes from the corresponding period a year before.

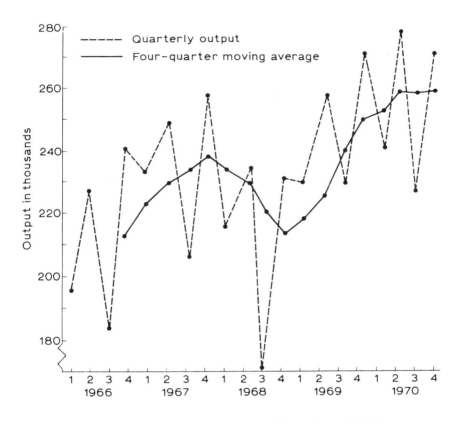

Figure 2.1   Output of new cars in Ruritania, 1966–1970

Possible refinements include a correction factor to make allowance for the different numbers of working days in each month, for a 12-month moving average. A moving average can also be used where there are large and irregular variations, for example of annual profits, to show a trend more clearly; here a ten-year cycle would be suitable.

Weighted means

A boy gains 43 % in an algebra exam and 67 % in geometry. What is his average mark, if algebra is to be counted as twice as important as geometry? The method is to count the algebra mark twice, and then to divide by three, giving 51 %. The algebra is said to be given a **weighting** of two, and the effect is to bias the mean away from the simple mean of 55 % towards the algebra mark. The calculation is expressed mathematically as

$$m = \frac{2x_1 + x_2}{2 + 1}$$

where x_1 is the algebra mark and x_2 the geometry. More generally,

$$m = \frac{w_1 x_1 + w_2 x_2 + w_3 x_3 \ldots}{w_1 + w_2 + w_3 \ldots} \tag{2.1}$$

where the w's are the weightings.

An **index number** such as a cost-of-living index is a further example of a weighted mean. Here the typical prices of the ordinary requirements of life are each given a weighting according to the amount needed, and the result is a measure of the change in overall cost.

A simple mechanical analogy often helps in calculating the weighted means of two quantities. Imagine a metre rule (assumed weightless) with 100 g weights hung at the 43 cm and 67 cm points to represent the marks in algebra and geometry. Then the arithmetic mean is at the point of balance between the two: 55 cm. For a weighted mean as before, the algebra mark is represented by *two* 100 g weights. The point of balance must now be twice as far from the 67 mark as the 43, by the laws of mechanics. The weighted mean is thus 51 marks, one-third of the way between 43 and 67. The same principle can be applied in finding an overall mean for two groups each with a known mean. For example, find the mean height of a class of children made up of 24 boys of mean height 159 cm and 8 girls of mean height 151 cm. Here there are three times as many boys as girls, so that the 'balance point' is three times as far from 151 as from 159, that is, 157 cm.

Two other means

A different type of mean, which is useful for calculating intermediate points in a time series and occasionally in other places, is the **geometric mean**. For example, if a country had four million inhabitants in 1800 and nine million in 1900, what might one expect it to have been in 1850? (Approximately, of course, assuming somewhat idealised conditions.) Not the arithmetic mean of $6\frac{1}{2}$ million, because the growth rate will be greater towards the end of the century when the population is greater. (The growth curve is in fact the exponential curve again.) The answer is six million, so that the increase is fifty per cent in each of the 50-year periods. The geometric mean is best calculated by taking the square root of the product—in this case, working in millions, the square root of four times nine. In order to calculate other intermediate points on the exponential curve, other methods are necessary, as mentioned at the end of chapter 1: the logarithm of the variable is directly proportional to the time.

A further mean, the **harmonic mean,** is only rarely used in statistical work. It is the number whose reciprocal is the arithmetic mean of the reciprocals of the variates.

Calculation of a mean—the Σ notation

The rest of this chapter is concerned solely with the arithmetic mean, because it alone is a suitable representative figure to use in the various statistical tests described in this book. The basic definition has already been given, on page 15, and the next stage is to develop methods of calculation. In this we shall use a form of notation which may be unfamiliar to some readers. For a distribution of n variates, labelled $x_1, x_2, x_3 \ldots x_n$, it is necessary, in finding the mean, to work out the sum of the variates, and so a shorter way of writing $x_1 + x_2 + x_3 + \ldots + x_n$ is used. This is $\sum_{1}^{n} x_r$, where the Greek capital sigma means 'take the sum of' and the 1 and n give the limits; x is used to denote a typical variate. The limits are usually obvious, and are omitted except where clarity requires them; the subscript r is often omitted also. The mean, therefore, defined as the sum of the variates divided by the number of them, is calculated from the formula

$$(1/n)\Sigma x. \qquad (2.2)$$

The 'bar' over the x is the standard way of denoting a mean value, and in speech it is read as 'x bar'. The letter m can also be used, as it has been already for a weighted mean, and μ (Greek mu) will be used later.

The sigma notation is analogous to the integral sign; the difference is that it is used with discrete rather than continuous variables.

Calculation of a mean—the use of a working mean

What is the mean of 177, 171, 178 and 174? If you have actually worked it out before continuing to read, it is likely that you have used a method like this: the figure 170 is common to all, so it is only necessary to find the mean of 7, 1, 8 and 4, which is 5, and the answer is 175. The same method could be used for 173, 166, 178 and 167, counting -4 and -3 for the two figures less than 170. The mean is then 171. This is the basis of the most usual method of calculating the mean of a distribution, the reference figure, which is 170 in these two examples, being known as the **working mean**. The term is not an accurate one, but is descriptive at least of intention if not of fact. Alternatives are assumed mean, false zero, arbitrary origin, or datum.

The mathematical justification for using a working mean is as follows. The variates are $x_1, x_2 \ldots x_n$, the working mean is x_0, and the difference between variate and working mean, $x_r - x_0$, is written d_r or simply d. This is known as the **deviation**. (Sometimes deviations from a working mean will be used in the same calculation as deviations from the actual mean, and it is necessary to keep track of which is which.) Then, from the definition of a mean (equation 2.2),

$$\bar{x} = (1/n)\Sigma x$$

$$= (1/n)\Sigma(x_0 + d)$$

$$= (1/n)(\Sigma x_0 + \Sigma d)$$

$$= (1/n)(nx_0 + \Sigma d)$$

$$\therefore \bar{x} = x_0 + (1/n)\Sigma d \qquad (2.3)$$

Notice that Σx_0 means count up as many of the x_0's as there are, and there are n of them.

A typical calculation is set out as follows.

x	d		$d+1$	
261	1		2	
244		-16		-15
253		-7		-6
272	12		13	
249		-11		-10
267	7		8	
262	2		3	
$x_0 = \overline{260}$	$\overline{22}$	-34	$\overline{26}$	-31
$n = \quad 7$	$\Sigma d = -12$		$\Sigma(d+1) = -5$	

Check: $\Sigma(d+1) = \Sigma d + n$
$$-5 = -12 + 7 \;\checkmark$$
$$\therefore \bar{x} = x_0 + (1/n)\Sigma d$$
$$= 260 + (1/7)(-12)$$
$$= \underline{258 \cdot 3}$$

The purpose of the third column is to make a check on the accuracy of the addition. In a simple example like this one cannot pretend that a check is necessary, but what *is* necessary is to establish at the outset that, wherever a simple arithmetical check is possible in a tabular calculation, then such a check is made. The reader who never makes mistakes can, of course, omit it, but lesser mortals should start forming the habit now. The algebraic formula justifying the method is given, and should not need any explanation.

The layout of the calculation should explain itself also, but some features may be pointed out. The working mean and the number of variates are written down, and likewise the formulae as they are used. The value of Σd is negative, and so the mean is less than the working mean. The answer is given to one place of decimals; whether this is the best choice must depend on the accuracy of the data and the future use of the answer. Finally, although a check is made, it does not check everything, and therefore you should take particular care to avoid or find errors in the remaining parts: in transcribing the variates, in working out the deviations, and in the final section.

In simple cases, it is possible to work out the mean without writing down anything at all. In this example, you can find the deviations one by one by subtracting 260 from the variates, and keep a running total at the same time. The final figure of -12 is then divided by 7 and added to 260 as before. Exponents of the art of mathematical one-upmanship can profitably try this on their friends, but should take care that in the chosen example the deviations are fairly small; the list of variates can be of any length.

Calculation of a mean—grouped frequency distributions

Often the number of variates is too large to be taken one by one, and so the distribution is divided into groups as described in chapter 1. Except in the very simple case where equal values of the same discrete variable are grouped, the process involves some loss of information, and the calculation which follows will only give an approximate value for the mean. The error is not usually a serious one, mainly because the many small errors of which it is made up tend to cancel each other out. It is not essential to understand why this happens, and therefore why the results may be taken as reliable—at a first reading at least you may prefer to skip the next three paragraphs. But the explanation is not difficult, and is worth coming back to later.

We shall take as an example the distribution shown in the histogram of figure 1.1, for which the figures are:

Ranges of marks:	0–9	10–19	20–29	30–39	40–49	50–59	60–69	70–79	80–89	90–100
Frequency:	2	3	8	11	15	34	53	33	13	3

The variable is discrete, and the variates are grouped with a class interval of 10. There are 34 variates in the 50–59 class, and in the calculation it is assumed that these lie evenly on each side of the class mid-point of 54·5. This will not in general be true, and for two reasons which should be clearly distinguished.

The first reason is that, because of the general shape of the distribution, it is likely that there were in the original figures more candidates in the 55–59 range than in the 50–54. The resulting error will underestimate the sum of the variates, and the same will happen in every group below the modal group. But for groups above the modal group, the sum of the variates will be overestimated, and so, provided that the distribution is not too asymmetrical, the overall error is small and without bias—that is, it does not tend consistently to overestimate or underestimate. If the distribution is an asymmetrical one, the errors may be minimised by keeping the class interval small.

The other reason for error may be seen in the modal group, where the error mentioned above does not occur. The class mid-point of 64·5 is taken as the class mean, but because of the nature of the variates, this will not in general be so—in fact in this case it could not be exactly so, because there is an odd number of variates. What can be said, however, is that there is no reason to suppose that it should be less than 64·5 rather than more. So again, there is no bias, and the errors tend to cancel each other out.

Now we come to the calculation itself. The number of variates in each class, the frequency, is labelled f. The formula for the mean is

$$\bar{x} = x_0 + c\frac{\Sigma fd}{\Sigma f},\tag{2.4}$$

in which there are three changes from the formula previously used. The number of variates n is replaced by Σf which means the same thing; the two are interchangeable. Σd is replaced by Σfd, but this is not really a difference at all—it is still the sum of the deviations, but calculated in a new way. The introduction of c, the size of the class interval, is necessary because the d's are measured in units of the class interval rather than in the original units—here, in tens of marks rather than in single marks. Where the class interval is 10 this refinement makes little difference, but if it is 5 it represents quite a saving of effort.

x	f	d	fd	$d+1$	$f(d+1)$
0–9	2	-5	-10	-4	-8
10–19	3	-4	-12	-3	-9
20–29	8	-3	-24	-2	-16
30–39	11	-2	-22	-1	-11
40–49	15	-1	-15	0	$\overline{-44}$
50–59	34	0	$\overline{-83}$	1	$\overline{34}$
60–69	53	1	$\overline{53}$	2	106
70–79	33	2	66	3	99
80–89	13	3	39	4	52
90–100	3	4	12	5	15

$$x_0 = 54\cdot5 \qquad \overline{175} \qquad \overline{170} \qquad \overline{306}$$
$$c = 10 \qquad = \Sigma f \quad \Sigma fd \quad = \overline{87} \qquad \Sigma f(d+1) = 262$$

Check: $\Sigma f(d+1) = \Sigma fd + \Sigma f$
$$262 = 87 + 175 \;\checkmark$$

$\therefore \bar{x} = x_0 + c\Sigma fd/\Sigma f$
$$= 54\cdot5 + 10 \times 87/175 = \underline{59\cdot5}.$$

The only other new feature is that the working mean has to be taken at one of the class mid-points in order to make the d's whole numbers. If the distribution has unequal class intervals, some fractions may occur in the d column; in these cases, care is necessary. With an unbounded class, as in a top age-group, an arbitrary choice of mid-point has to be made.

Exercises

(Further exercises in the calculation of means will be found at the end of chapter 3.)

1 A boy's marks in history and geography are 46 and 73 respectively. Find the mean mark, (a) if history is given a weighting of 2, and (b) if history is given a weighting 4 and geography 5.

2 Find the mean of the following marks: 61, 47, 52, 56, 59, 50, 49, 57, 55, 64, 52, 56, 60, 48, 46, 67, 53, 57, 55, 54.

3 Find the mean of the following ages, which are in years and completed months: 13·7, 15·0, 14·7, 14·1, 14·6, 14·3, 13·8, 13·11, 14·11, 15·2, 14·7, 14·6, 14·11, 14·3, 14·0.

4 In 600 throws of a die the following numbers of scores were recorded. Calculate the mean score.

Score:	1	2	3	4	5	6
Number:	107	89	95	99	104	106

5 Eight coins were thrown 1000 times, and the numbers of heads recorded each time. Calculate the mean number of heads per throw.

Number of heads:	0	1	2	3	4	5	6	7	8
Frequency:		3	27	110	207	291	215	109	33	5

6 Find the mean height of the following group of children: the heights were measured to the nearest centimetre and then grouped.

Heights (cm)	110–119	120–129	130–139	140–149	150–159	160–169
Numbers	1	19	48	95	31	6

7 A cost-of-living index is based on the changes of four separate indices for food, fuel, rents and clothing. These are weighted according to the table below, which shows also the values of the four indices for two successive years. Find the percentage change in the cost-of-living index.

	Weighting	Index, first year	Index, second year
Food	45	113	121
Fuel	15	127	123
Rents	30	108	119
Clothing	10	110	131

8 The failure rates f of electronic equipment over equal successive intervals of time are shown in the following table against the central value t of each interval:

t:	1	2	3	4	5	6	7	8	9	10
f:	0·012	0·019	0·048	0·028	0·029	0·063	0·033	0·054	0·063	0·058

t:	11	12	13	14	15	16	17	18	19	20
f:	0·064	0·078	0·068	0·074	0·098	0·086	0·076	0·113	0·106	0·081

Use a 3-point moving average to smooth these results. Plot the smoothed results on graph paper and draw by eye a straight line to fit the points. Estimate the failure rates for the intervals centred at $t = 21$ and at $t = 22$. (O & C)

9 Prove that, if $x \neq y$. and if m, g, and h are respectively the arithmetic, geometric and harmonic means of x and y, then $m > g > h$.

Chapter 3

The standard deviation

The only way to convey all the information contained in a set of statistics is to write down every single figure. But this is usually impracticable because of the labour involved; and even if it were possible, it would not give an easily readable result. One way of simplifying the figures, as we found in the last chapter, is to take the mean, but in doing so one aspect of the overall situation is left completely out of account. This is the range of values covered by the distribution, and the relative numbers of each value: in a word, the **dispersion** of the distribution. The sketches below show something of the variety of shapes that distributions with the same mean can have.

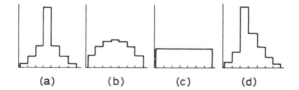

Figure 3.1   Distributions which have the same mean

In this chapter we shall use a number of ways of expressing these differences quantitatively. No single figure can describe completely the shape and range of a distribution, and so some information about it will still be lost; for example, figure (d) is a **skew** or asymmetrical distribution, and a separate measure is required for skewness. Of the various measures of dispersion, one, the standard deviation, is far more useful than all the others. The standard deviation marks the point at which most people's knowledge of statistics comes to an abrupt halt, and the specialised work of the statistician begins. It is used, directly or implicitly, in every one of the statistical tests in this book. Other measures, however, are sometimes used because of their relative simplicity, and a brief mention of them must be made.

First and simplest is the **range**. This is the difference between the largest and smallest values of the variable. It is very simple to calculate, and gives a useful first impression of the dispersion. But for detailed analysis its value is small, since it takes account of only two members of the distribution. All four of the sketches above, for example, could have the same range.

Next is a measure of the range covered by the central 50% of the distribution. This makes use of a concept similar to the median, called a **quartile**. The median is the middle variate taken in size order; the lower and upper quartiles are the variates one-quarter and three-quarters of the way through the order. As with the median, the points are found from the formulae $\frac{1}{4}(n+1)$ and $\frac{3}{4}(n+1)$ rather than $\frac{1}{4}n$ and $\frac{3}{4}n$, out of n observations, but some rounding off is justifiable. For a grouped distribution, the quartiles may be found either from a cumulative frequency curve or by an arithmetical method using simple proportion; for details see the section on the median, page 14. The measure of dispersion is half the difference between the quartiles, the **semi–interquartile range**. A briefer term meaning the same thing is the **quartile deviation**.

As an alternative to the quartiles, the outer sextiles may be used. The sextiles are the points dividing the distribution into sixths, and the outer ones are, of course, the $\frac{1}{6}$ and $\frac{5}{6}$ points, with the central two-thirds of the distribution between them. The **semi-intersextile range** (the 'outer' being understood) so defined is useful because in many types of distribution it gives a rough estimate of the standard deviation. Both of these measures of dispersion are much better than the range, but neither takes account of the length of the 'tails' of the distribution.

These three measures of dispersion are illustrated, using the typical cumulative frequency curve, in figure 3.2. Other similar measures are occasionally used: the general term for a variate dividing a distribution at a given point is a **quantile**, and one particular type is the **percentile**, whose name explains itself.

In order that each member of the distribution be represented in the measure of dispersion, we can use the mean deviation from the mean. The calculation is simple if the mean is a whole number, and, if it is not, it can be rounded off to one without serious error. In adding up the 'd' column, of course, all the signs are taken as positive. The process is open to objection on the grounds that the deviations ought to be measured from the median, not the mean, if the sum of the deviations is to be a

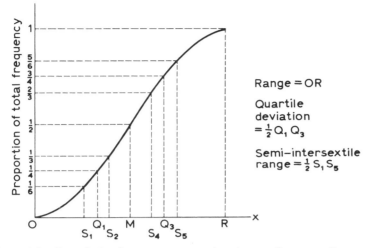

Figure 3.2   Cumulative frequency curve showing median, quartiles, etc.

minimum. However, we need not pursue the point since the mean deviation is rarely used in practice. The reason for this may be seen by comparing it with the standard deviation, to which we now turn.

The standard deviation is also an average deviation from the mean, but an average calculated in a rather special way. Suppose you were asked to find the average size of three carpets, 1·5 m square, 2·1 m square and 3·9 m square respectively. One way would be to add the three figures together and divide by three, giving 2·5 m. But clearly this would not be very sensible, as the quantity which should be averaged is the area, represented by the *square* of the length of the side. The steps in the calculation are then as follows:

(i) Square each measurement, giving 2·25, 4·41 and 15·21.

(ii) Add, giving 21·87.

(iii) Divide by the number of measurements (3), giving 7·29.

(iv) Take the square root of the result, giving 2·7.

The average size is therefore that of a carpet 2·7 m square, not 2·5.

This is exactly the process used in the calculation of a standard deviation, with the lengths of the carpet sides being replaced by the deviations from the mean. In mathematical language, the standard deviation (abbreviated to SD from now on) is the square root of the mean squared deviation from the mean, or more briefly the root mean square deviation. In symbols it is

$$s = \sqrt{\left(\frac{\Sigma d^2}{n}\right)}. \tag{3.1}$$

B

For the moment, the letter s will be used for the SD; later on, σ (the Greek small sigma) will also be used.

The effect of finding the average in this way is that the larger deviations are given more 'weight'. The observations in the tails of a distribution therefore count proportionally more in the size of the SD than they do in the mean deviation, and it is possible to justify the use of the squaring process on these grounds only. Consider these two distributions, which represent the errors in 16 measurements made by each of two surveyors.

X: 2 −2 0 1   2 −1 2 0 −1 −1 0 2 −2 −1 0 −1
Y: 0 −1 3 0 −2   0 0 1 −1 −4 0 1 −1   0 3   1

Each has a mean of zero and the same mean deviation. But which one is the better surveyor? Arguably X, since he consistently keeps within ±2 of the correct value. By contrast Y, although he reads correctly on more occasions, sometimes makes more serious errors than X ever does. The relative importance of a given deviation from the mean is usually more than in proportion to its own size, and squaring is the easiest way to take account of the fact. For these distributions, the SD's are in fact 1·37 and 1·66 respectively.

The SD has practical advantages also. The root mean square deviation is a minimum when deviations are measured from the mean, not the median; the squaring process deals with negative deviations automatically, since all the squares must be positive; it is susceptible to certain useful mathematical devices in analysis; and finally it is more reliable in use, though the theoretical justification for this is too advanced for an introductory textbook. To sum up, the word 'standard' in its name is fully justified.

A related quantity, the **variance**, is often used. This is simply the square of the SD, s^2 or σ^2. It avoids the use of the square root sign, and is more convenient in some of the formulae of later chapters. Often the two are used in different parts of the same calculation, and care is needed in keeping to the right figures.

In comparing two distributions, the **coefficient of variation** is sometimes useful: this is the SD expressed as a percentage (or sometimes a fraction) of the mean. Since the SD is measured in the same units as the original figures (if you are doubtful of this, go back to the carpet example), the coefficient of variation is dimensionless. It can therefore be used to compare two distributions measured in different units, for example the heights and weights of a group of adults.

Calculation of a standard deviation

This is set out in tabular form. The method using a working mean will be described later; until then, the deviations must be measured from the actual mean, which is only convenient when the mean is a whole number. As in the calculation of a mean, a check column is used, and an example will make the method clear.

Example 1 Find the mean and SD of the distribution given in the first column below.

x	d	d^2	$d+1$	$(d+1)^2$
19	-8	64	-7	49
22	-5	25	-4	16
25	-2	4	-1	1
25	-2	4	-1	1
28	1	1	2	4
29	2	4	3	9
34	7	49	8	64
34	7	49	8	64
$\overline{216}$	$\overline{0}$	$\overline{200}$	$\overline{8}$	$\overline{208}$
$= \Sigma x$	$= \Sigma d$	$= \Sigma d^2$	$= \Sigma(d+1)$	$= \Sigma(d+1)^2$

The first stage of the calculation is to find the mean, which is $(1/n)\Sigma x$, $= 216/8 = 27$. (A working mean can be used for this part of the calculation if desired.) Then the d column is completed, and as a check on the mean Σd should be zero. The other three columns are next completed and summed. The check on Σd^2 is then made, using the identity

$$\Sigma(d+1)^2 = \Sigma d^2 + \Sigma(d+1) + \Sigma d,$$

which follows from the fact that $(d+1)^2 = d^2 + (d+1) + d$. In this example, the three terms are 200, 8 and 0, and the sum is 208 as it should be. Hence

$$s = \sqrt{(\Sigma d^2/n)} = \sqrt{(200/8)} = 5.$$

The two quantities $\bar{x} = 27$ and $s = 5$ give a description of the distribution sufficient for many purposes.

Grouped frequency method of calculation

The general idea is much the same as when this method was used for the mean of a large distribution. The letter f appears before the d^2, indicating, not a different formula, but a different method of computation, and the final result is multiplied by c, the class interval, to allow for the fact that the d's are measured in units of c. The formula for the SD is then either of the two forms

$$s = c\sqrt{(\Sigma fd^2/n)} = c\sqrt{(\Sigma fd^2/\Sigma f)}. \qquad (3.2)$$

Example 2 This example is somewhat artificial, in that the mean is arranged to be equal to one of the class mid-marks, 34·5; it is included only to show the principle of this type of calculation. In practice, a working mean method would be used. The distribution is the grouped marks of 200 candidates in an examination, and the mean and SD are to be found.

x	f	d	fd	fd^2	$d+1$	$f(d+1)$	$f(d+1)^2$
0–9	3	−3	−9	27	−2	−6	12
10–19	13	−2	−26	52	−1	−13	13
20–29	51	−1	−51	51	0	−19	
30–39	67	0	−86		1	67	67
40–49	47	1	47	47	2	94	188
50–59	18	2	36	72	3	54	162
60–69	1	3	3	9	4	4	16
	200		86	258		219	458

$x_0 = 34\cdot5 = \Sigma f \qquad \Sigma fd = 0 \qquad = \Sigma fd^2 \quad \Sigma f(d+1) = 200 \quad = \Sigma f(d+1)^2$

Checks: $\Sigma fd + \Sigma f = 0 + 200 = 200 = \Sigma f(d+1)$ ✓

$\qquad \Sigma fd^2 + \Sigma f(d+1) + \Sigma fd = 258 + 200 + 0 = 458 = \Sigma f(d+1)^2$ ✓

The mean is therefore equal to x_0, 34·5, and the SD is

$$s = c\sqrt{(\Sigma fd^2/\Sigma f)} = 10\sqrt{(258/200)} = 10\sqrt{1\cdot29} = 11\cdot3.$$

In a relatively large distribution such as this, a simple approximate check of the SD is possible—the only condition is that the distribution must tail off reasonably smoothly at both ends, or in other words must be roughly of the typical bell-shape pattern. The check is that the semi-intersextile range of such distributions is almost the same as the SD. Here the outer sextiles, as found from a cumulative frequency graph, are at approximately 23 and 47 marks, so that the semi-intersextile range is 12. A check as accurate as this should not always be expected, but a difference of over 15% or so should be looked at very closely.

As an alternative check, Table 1 (page 177) may be used, if the number in the sample is 17 or less: the range of the sample, multiplied by the number in the table, gives an approximate figure for the SD. For larger samples, the SD will usually be between a sixth and a quarter of the range.

Calculation using a working mean

In order to use a working mean, it is necessary first to find a relationship between the sum of the squares of the deviations from the actual mean (which we want) and the sum of the squares of the deviations from some other given value. This relationship will form the basis of a number of other formulae which we shall use later on, and is important.

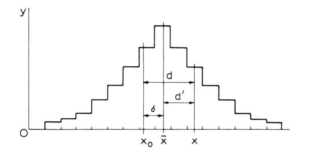

Figure 3.3   Typical distribution, for working mean calculation

Let d' be the deviation from the true mean $\bar{x}$ of a variate x, and let d be the deviation from an arbitrary value x_0, which may be a working mean or, in certain cases, zero. The symbol δ (Greek delta) is used for the difference between $\bar{x}$ and x_0; it does not matter which of these two is the greater since only $\delta^2 = (\bar{x} - x_0)^2$ appears in the final formula. The problem is to find $\Sigma d'^2$ without using the individual values of d'. Since $d = d' + \delta$,

$$\Sigma d^2 = \Sigma(d' + \delta)^2$$
$$= \Sigma d'^2 + \Sigma 2d'\delta + \Sigma \delta^2 \quad \text{(i)}$$
$$= \Sigma d'^2 + 2\delta\Sigma d' + n\delta^2 \quad \text{(ii), (iii)}$$
$$= \Sigma d'^2 + n\delta^2$$
$$\therefore \Sigma d'^2 = \Sigma d^2 - n\delta^2$$
$$\therefore \Sigma d'^2 = \Sigma d^2 - n(\bar{x} - x_0)^2 \quad (3.3)$$

Before we use this result, the steps in the derivation need explanation, and they are important enough to be worth thinking through.

(i) The terms governed by the Σ sign can be split up, in an analogous way to the integration process for the sum of two quantities: we are, in fact, only adding up the same terms in a different order.

(ii) A constant like 2 or δ can be taken across a Σ sign: since for example $2 \times 1^2 + 2 \times 2^2 + 2 \times 3^2 = 2(1^2 + 2^2 + 3^2)$, $\Sigma(2x^2) = 2\Sigma x^2$. Also, $\Sigma\delta^2$ means add up as many of the constants δ^2 as there are, and there are n of them.

(iii) Since d' is a deviation from an actual mean, $\Sigma d'$ is zero.

The SD can now be worked out:

$$s^2 = \Sigma d'^2/n = \Sigma d^2/n - (\bar{x} - x_0)^2.$$

From equation 2.3, $\bar{x} = x_0 + (1/n)\Sigma d$, and so

$$s^2 = \Sigma d^2/n - (\Sigma d/n)^2, \quad \text{and} \quad s = \sqrt{\{\Sigma d^2/n - (\Sigma d/n)^2\}} \qquad (3.4)$$

This is probably the formula you will have to use most frequently, and after you have had some practice with it, it should be memorised. The second term in the formula represents a 'correction factor' to be subtracted, which is made necessary because the d's are measured from a point other than the mean. If the working mean is chosen to be fairly close to the actual mean, this term can be made quite small (typically of the order of 1% of the other). If this is the case, it will not need to be evaluated to a high order of accuracy.

Two points may be noted in passing.

(i) The correction factor is always positive, since it is a square. In the formula for s^2, s is a constant for any given distribution, so that $\Sigma d^2/n$ is a minimum when $(\Sigma d/n)^2 = 0$, that is, when $\Sigma d = 0$. But this occurs when deviations are measured from the actual mean. In other words, the sum of the squares of the deviations from the mean is less than the sum of the squares of the deviations from any other value. The mean could, in fact, be defined as the point from which the sum of the squared deviations is a minimum. These properties will be needed on more than one occasion later on.

(ii) When the working mean is zero, the formula becomes:

$$s^2 = \Sigma x^2/n - (\Sigma x/n)^2 = \Sigma x^2/n - \bar{x}^2. \qquad (3.5)$$

Since the original observations x are usually much larger than the deviations from a working mean, this form is not suitable for hand calculations. But when a calculating machine is available, size of numbers is not important, and it is an advantage to be able to use the data as it is without having to calculate deviations first. The alternative form is also sometimes useful in mathematical analysis.

If you have found the derivation of these equations difficult, it may help to make the process clearer if we derive equation 3.5 directly, again using equation 3.1 as a starting point, as follows.

$$s^2 = (1/n)\Sigma d^2$$
$$= (1/n)\Sigma(x - \bar{x})^2$$
$$= (1/n)\Sigma x^2 - 2\bar{x}(1/n)\Sigma x + (1/n)\Sigma \bar{x}^2$$
$$= (1/n)\Sigma x^2 - 2\bar{x}^2 + \bar{x}^2$$
$$= (1/n)\Sigma x^2 - \bar{x}^2$$
$$= (1/n)\Sigma x^2 - (\Sigma x/n)^2.$$

Since the SD is determined solely by the deviations from the mean, its value is not affected by adding any chosen number to each variate. Thus the 'zero' can be altered at will, or in other words, the x in the equation above may be replaced by the deviation from any desired value. This value will be the working mean, and we are back to equation 3.4 once more.

Example 3 Find the mean and SD for the distribution of x below.

x	d	d^2	$d+1$	$(d+1)^2$
33	-12	144	-11	121
36	-9	81	-8	64
38	-7	49	-6	36
43	-2	4	-1	1
45	0	0	1	1
49	4	16	5	25
49	4	16	5	25
53	8	64	9	81
54	9	81	10	100
56	11	121	12	144

$x_0 = 45 \quad \Sigma d = 6 \quad \Sigma d^2 = 576 \quad \Sigma(d+1) = 16 \quad \Sigma(d+1)^2 = 598$

Checks: $\Sigma d + n = 6 + 10 = 16 = \Sigma(d+1)$ ✓

$$\Sigma d^2 + \Sigma(d+1) + \Sigma d = 576 + 16 + 6 = 598 = \Sigma(d+1)^2 \text{ ✓}$$

$$\therefore s^2 = \Sigma d^2/n - (\Sigma d/n)^2 = 576/10 - (6/10)^2 = 57{\cdot}24$$

$$\therefore s = 7{\cdot}57, \quad \text{and} \quad \bar{x} = 45{\cdot}6.$$

Note that, as for the mean, the check here does not cover the subtraction to give the individual deviations, or the final stage. In the uncheckable parts, particular care should be taken to avoid errors. Perfectionists will repeat the calculation with another working mean.

Example 4 This uses the version of the formula suitable for grouped frequency calculations, which has already been given in its simple form (3.2) and now needs adapting for use with a working mean. It becomes

$$s = c\sqrt{\{\Sigma fd^2/\Sigma f - (\Sigma fd/\Sigma f)^2\}}. \tag{3.6}$$

The distribution represents the heights of 300 adults, measured to the nearest centimetre and then grouped with a class interval of 5 cm; the mean and SD are to be found.

x	f	d	fd	fd^2	$d+1$	$f(d+1)$	$f(d+1)^2$
150–154	6	-4	-24	96	-3	-18	54
155–159	29	-3	-87	261	-2	-58	116
160–164	54	-2	-108	216	-1	-54	54
165–169	68	-1	-68	68	0	-130	
170–174	70	0	-287		1	70	70
175–179	47	1	47	47	2	94	188
180–184	22	2	44	88	3	66	198
185–189	4	3	12	36	4	16	64
$\Sigma f = 300$			103	812		246	744

$x_0 = 172$ $\Sigma fd = -184$ $= \Sigma fd^2$ $\Sigma f(d+1) = 116$ $= \Sigma f(d+1)^2$

Checks: $\Sigma fd + \Sigma f = -184 + 300 = 116 = \Sigma f(d+1)$ ✓

$\quad\quad\quad \Sigma fd^2 + \Sigma f(d+1) + \Sigma fd = 812 + 116 - 184 = 744 = \Sigma f(d+1)^2$ ✓

$\quad\quad \therefore s = c\sqrt{\{\Sigma fd^2/\Sigma f - (\Sigma fd/\Sigma f)^2\}} = 5\sqrt{\{812/300 - (184/300)^2\}}$

$\quad\quad\quad\quad = 7\cdot63$ cm.

$\quad\quad \bar{x} = x_0 + c\Sigma fd/\Sigma f = 172 - 5 \times 184/300 = 168\cdot9$ cm.

We can now calculate the SD of any set of observations we meet. But such calculations are not ends in themselves, and you should spend no longer on the study of them than is necessary. They are for statistics what the arithmetical operations of addition, multiplication and so on are for mathematics: the preliminaries, and no more.

Exercises

In questions 1 to 14, the means and standard deviations are to be found; in 1 to 6 the means are round numbers, but after that the formulae using a working mean give the most economical method.

1 15, 17, 17, 19, 19, 21, 21, 23, 25.

2 5, 6, 8, 8, 11, 14, 14, 14.

3 1·5, 1·6, 1·7, 1·7, 1·8, 1·8, 1·9, 1·9, 1·9, 2·0, 2·0, 2·1, 2·1, 2·1, 2·2, 2·2, 2·3, 2·3, 2·4, 2·5.

4 15, 17, 18, 18, 19, 20, 21, 23, 24, 25.

5 x: 0   1   2   3   4   5   6   7
   f: 5   11   19   29   23   8   3   2

6 x: 0   1   2   3   4   5   6   7   8
   f: 3   32   117   215   266   220   109   33   5

7 36, 37, 38, 38, 39, 40, 41, 41, 42, 44.

8 0·003, 0·005, 0·008, 0·008, 0·010, 0·013, 0·014, 0·016, 0·017, 0·019.

9 51·3, 52·5, 51·8, 52·6, 51·8, 51·6, 52·5.

10 -12, -10, -9, -7, -6, -4, -3, -1, 1, 2.

11 x: 3   4   5   6   7   8   9
    f: 6   17   21   26   18   10   2

12 x: 2·3   2·4   2·5   2·6   2·7   2·8   2·9   3·0   3·1   3·2
    f: 2   8   16   25   38   43   33   19   11   5

13 Percentage marks (in groups of 10) of 500 examinees:

Marks:	1–10	11–	21–	31–	41–	51–	61–	71–	81–	91–100
Number of examinees:	17	34	33	65	75	86	89	56	32	13

14 Girths of trees in a single plantation, measured at a fixed height above the ground to the nearest centimetre.

x:	35–39	40–44	45–49	50–54	55–59	60–64	65–69
f:	17	56	98	63	38	25	3

15 For the data of questions 11 to 14, estimate the medians, quartiles and outer sextiles; note the comparative figures for the SD and the semi-intersextile range.

16 For the distribution 1, 2, 2, 3, 3, 3, 4, 4, 5, find the sum of the squared deviations from 0, 1, 2, 3, 4, 5 and 8. Notice that it is a minimum at 3, the mean of the distribution, and that it rises rapidly as the reference point moves away from 3.

17 A group of children took two examinations in English and Mathematics, both marked out of 100. In English the mean was 64 and the SD was 8; in Mathematics the figures were 48 and 16. How should the marks be combined into a single order if the subjects are judged to be of equal importance?

Chapter 4

Probability

The nature of probability

Any non-statistical mathematician who has picked up this book will, I hope, have felt himself quite at home in the first three chapters. They have been concerned with a tidy world, where precise formulae are applied to the data and churn out an answer whose value is in no doubt, except perhaps for some rounding off at the end. A mathematician's typical problem is that of the simple pendulum, a mass making small swings at the end of a string whose other end is fixed. He applies the laws of Newtonian mechanics and derives a formula; he applies this to the data, and makes a prediction of the time of swing. Logically it is all very satisfactory, and a good experimenter could give a fair verification of the answer. But in making such a prediction, many of the realities of the situation are ignored. The data is not accurately known, and the effects of air resistance (and other things also) are left out of account. In this case it happens that the errors are small, and so a good prediction can be made.

Alas for the pure mathematician, such tidy situations are the exception rather than the rule. Sometimes when simple methods fail, more refined ones can be used, as for example with objects moving through air at higher speeds, when air resistance has to be taken into account. But all too often, the data is so complex that to attempt even a simplified form of calculation is totally out of the question. Consider for example the spin of a coin.

The factors which determine the path of the coin through the air and its final resting place include the following: the height of the spinner's hand above the ground; the initial position of the coin; the strength, point of action and direction of the blow from the spinner's thumb; the mass and dimensions of the coin; the effects of air resistance; the wind speed and direction; and finally the nature of the surface on which the coin falls. Even if these were measurable, and in the ordinary situation most of them are not, the problem is much too difficult.

In order to deal with this situation, we use the concept of probability. This is an attempt to set down in a quantitative manner what is known to happen in practice, modified in some cases by considerations of symmetry.

The starting point of the procedure is an **experiment** whose outcome is measurable, but not precisely predictable on theoretical grounds because of its complexity. The experiment may be repeated any number of times, with the condition only that changes in the data must not be such as to affect the outcome in a predictable way. (Such repetitions are called **trials**.) For example, if a coin were pressed into a concave-convex shape, one could predict that it would come to rest convex side downwards more often than before. But if the coin were held at a different height above the ground, or if it were held tail-up rather than head-up, this would not make prediction of the result any easier.

Some types of experiment, and the spin of a coin is one of them, are such that it is possible to predict the result quantitatively on grounds of symmetry. It is not necessary that this should be so, and the first experiment we shall consider in detail is not a symmetrical one. Instead of a coin, an ordinary drawing pin is tossed up and comes to rest on a flat surface, either with its point upwards or resting partly on the point and partly on the head (a 'point' or a 'head'). It is quicker to throw a number at once, and quite fair provided that they are indistinguishable. The following figures show the number of 'heads' in 100 actual throws of 20 unused drawing pins taken from the same box.

6 11 9 8 10 / 8 8 5 12 8 / 4 8 7 10 6 / 5 8 6 9 7 /

10 9 8 7 11 / 5 9 7 8 7 / 6 4 4 9 4 / 4 8 7 12 12 /

5 9 10 5 12 / 8 8 8 9 6 / 11 9 7 5 9 / 8 8 8 9 5 /

5 8 8 5 9 / 4 9 7 11 8 / 7 9 9 4 13 / 9 12 5 8 9 /

10 10 11 8 10 / 6 11 4 6 5 / 13 10 7 9 5 / 10 6 13 8 8

It is easy to work out from these figures the proportion of heads that appeared when a given number of drawing pins were thrown. A few such figures are shown below.

Total number thrown:	20	100	500	1000	1500	2000
Number of heads:	6	44	200	386	581	794
Proportion of heads:	0·300	0·440	0·400	0·386	0·387	0·397

The theory of probability states that, if such an experiment were continued long enough, the variations in the figure for the proportion would become less and less, so that one would arrive at a single figure which represents the probability that a drawing pin tossed up will come to rest with its head upwards. Put more formally, the proportion of heads converges to a limit, called the probability, as the number of throws becomes very large.

The figures above do not, and could not, prove this, but they do indicate at least that it is plausible, the probability being about 0·4. To put it another way, the increase in heads which happened to occur in the last 500 throws *might* have been a systematic change rather than a perfectly ordinary one; all we are saying is that such a thing is contrary to common experience. If it did happen because, say, some of the drawing pins were bent in the course of throwing them, this would be contrary to our original assumption that there must be no changes in the data which could affect the outcome in any way.

The definition of probability in mathematical form is

$$p = \underset{n \to \infty}{\text{Lt}}\left(\frac{x}{n}\right)$$

where x is the number of occurrences of a particular result (an **event**) in n trials of the experiment. As a definition, this is open to the objection that it presupposes that such a limit exists. It is perhaps better described as an axiom: it is in accordance with common sense, it works in practice and there appears to be no way of giving a direct proof of it. The parallel lines axiom, on which the subject of Euclidean geometry is built, is of similar status.

It follows from the definition that probability is measured on a scale from zero (if $x = 0$) to 1 (if $x = n$). Zero represents impossibility—the chance that a drawing pin falling on to a hard surface will come to rest balanced on its point. One represents certainty—the probability that a double-headed coin will show a head after spinning. It also follows that, if the probabilities of the different possible outcomes of an experiment are $p_1, p_2, p_3 \ldots$, then $\Sigma p = 1$.

The definition is also open to practical objection on the grounds that one can never use it to determine an exact numerical value for a probability. Its usefulness is really in setting up a model which (modified in some respect or other) may be used to represent real situations.

It can be modified, first, by considering a large number of trials rather than an infinite number, as we have just done with the drawing pins. This, of course, gives an approximate value for the probability.

It can be modified, secondly, by considerations of symmetry, and in many cases this is both possible and useful. Suppose we have a well-constructed ordinary six-sided die, and throw it a very large number of times. Since the die is symmetrical, we have no reason to expect that there will be either more or less 6's than any other given score: that is, although there might well be a small difference between the two numbers, we have no reason for expecting one rather than the other to be the larger. (This, of course, is not true for an asymmetrical case.) Thus, in an infinite series of trials, the proportions of each score may be expected to tend to the same limit (i.e. probability), and this must be 1/6 for each.

This may seem a long way round to a rather obvious conclusion. But if the method of reasoning used here is taken as typical, some of the difficult points of later work may be overcome more easily.

The frequency definition interpreted in this way leads on naturally to an alternative definition of probability, as follows. If there are n possible results of an experiment, all equally likely on grounds of symmetry, then the probability of any one of a given group of m out of these n results is equal to m/n. Used as a first definition, this is open to objection because the phrase 'equally likely' implies some understanding of probability itself, which we are trying to define. Here we interpret the phrase to mean that the limits of the proportions of each of the n results should be equal on grounds of symmetry.

The definition of probability is not an easy one to make rigorous. But whatever precise definition is adopted, there is no disagreement about the rules to be followed in using it.

Note that we have not attempted to attach any meaning to a statement such as 'the probability that Gambler's Folly will win the Derby is $\frac{1}{3}$'. From our definition, if there is no experiment even of a theoretical nature which could determine the probability on a frequency basis, we cannot give a numerical value to it. There are, of course, other and wider definitions, but they would take us into difficult and controversial fields of study.

Probability distributions

The six probabilities for a die, and the two for a drawing pin, may be represented in diagram form as follows—a way of showing all the probabilities together like this is called a **probability distribution**. In the case of the drawing pin, the probabilities are, of course, approximate.

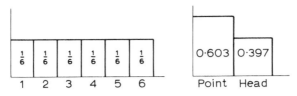

Figure 4.1   Probability distributions for die and drawing pin

In each case the probabilities are shown by areas, and the total area is therefore 1. Each is a probability distribution for a discrete variable: in the case of the drawing pin the variable has yet to be defined numerically, and it could be, say, 0 for a point and 1 for a head. But we can also have a probability distribution for a continuous variable.

Suppose we have a die of exactly cylindrical shape, and we mark a scale, say 0 to 1, round the complete circumference of one end. Then, if the die is rolled on a flat surface, we may take as the 'score' the value on the scale at the one point on it where the cylinder is in contact with the surface. The score may take any value between 0 and 1 on a continuous scale, with a similar probability basis to that of the six-sided die because the cylinder is also symmetrical. This is an idealised experiment, intended to give a picture of the kind of result more usually found in other ways, and so the practical difficulties of measurement and so on need not concern us. The probability distribution for the experiment looks like the left-hand part of figure 4.2.

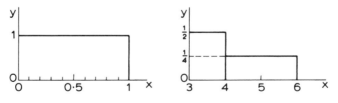

Figure 4.2   Probability distributions for the cylindrical die

Note that, in this case, we cannot speak meaningfully of 'the probability of a score 0·6': we have to specify a finite range of scores, such as 0·595 to 0·605, for which the probability is 0·01, the area between these limits. In general, if we call the score x, then the probability of a score in the range x to $x + \delta x$ is equal to $y\delta x$, where the height y of the block is the **probability density**. (Here $y = 1$ throughout.)

This imaginary experiment is useful in a number of ways. First, it generates a number (on a continuous scale) selected randomly, which in this case has a uniform probability distribution between 0 and 1. We shall call a number like this a **continuous random variable**. In the next paragraph we shall consider a random variable with a non-uniform distribution—a random variable may, in fact, have any distribution of finite density between any limits. The word 'random', by the way, is used here more or less in the sense in which it is used in ordinary speech, but to be strictly accurate its meaning is given by defining the random variable itself, as a variable which possesses a probability distribution.

Second, we can modify the distribution simply by altering the scale used. Suppose, for example, we mark one half of the circumference with a linear scale from 3 to 4, and the other half with a linear scale from 4 to 6. The probability distribution will then look like the right-hand part of figure 4.2. The score x is again a continuous random variable, but not all equal ranges of values have the same probability. The probability is still given by the quantity $y\delta x$ as above, as you should be able to check using a pair of numerical values such as 5 and 5·5. We could also develop probability distributions in which y varied continuously (rather than in steps) by using a non-linear scale round the circumference, for example a logarithmic scale: the value of y (as a function of x) would then be a **probability density function**. (These are considered in more detail in chapter 14.)

Third, we can use the cylinder to set up any required form of probability distribution for a discrete variable. If we divide the circumference into six equal parts, we have the analogy of a six-sided die; if we make the six parts not all equal, we simulate a biased die; if we use a non-linear scale of suitable form, and split it up into a number of sections, we can produce a distribution to order. This, again, is not particularly important in itself, but it is a useful model for comparing a probability distribution (developed mathematically for a continuous variable) with an actual distribution which will normally be for a discrete variable. An example will show the method.

Suppose we have a probability distribution which is a curved shape, part of which is shown in figure 4.3. We assume that the curve is worked out for x as a continuous variable, and we require the comparable distribution of X, a discrete variable, which takes integral values only.

We first have to decide the limits of x which correspond with each value of X, for example $X = 4$. If we set it to the right of the 4 mark, that is $4 < x < 5$, then the block for $X = -4$ must be $-4 < x < -3$, which seems inconsistent—if the distribution of x were symmetrical, that for X would not be. If we try to improve matters by offsetting negative values to the left, the two parts do not meet properly in the middle. The

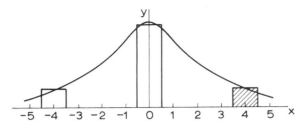

Figure 4.3   Comparison of discrete and continuous probability distributions

solution of the problem is to take the limits for $X = 4$ to be $3\frac{1}{2} < x < 4\frac{1}{2}$. This is rather like the method of measuring lengths to the nearest centimetre, where any measurement in the range $3\frac{1}{2}$ to $4\frac{1}{2}$ cm is counted as 4 cm. Thus $X = 0$ will correspond with $-\frac{1}{2} < x < \frac{1}{2}$, and $X = -4$ gives $-4\frac{1}{2} < x < -3\frac{1}{2}$; all is now neatly symmetrical.

We can also carry out the procedure in a reverse order—for example, to set up a continuous probability distribution to represent the scores on a die. The continuous scale in this case would go from $x = \frac{1}{2}$ to $x = 6\frac{1}{2}$, so that score 1 would correspond with $\frac{1}{2} < x < 1\frac{1}{2}$, and so on.

Quite often we require the probability that a random variable is greater than some specified value. Suppose we require the probability that $X \geqslant 4$; this corresponds with $x > 3\frac{1}{2}$, not $x > 4$. Similarly, $X > 4$ corresponds with $x > 4\frac{1}{2}$. The difference, usually an odd half one way or the other, is known as a **correction for continuity**.

The discrete random variable

With the rolling cylinder, we generated a continuous random variable which could have a variety of probability distributions. We have also considered some discrete distributions, and these likewise are probability distributions for a *discrete* random variable. The result of throwing a drawing pin, for example, is a random variable which takes the value 0 (understood to mean a 'point') with probability about 0·6 for the type used, and value 1 (a 'head') with probability about 0·4. Notice that 'random' does not imply that the probabilities of all possible values are equal, though this will often be the case. In general, the result of any physical experiment is a random variable, except in the rare case of a result which for some special reason (such as if it is reached by counting as opposed to measuring) can be expressed with absolute precision. The concept is an important one, and will be much used later.

A probability distribution, for the moment, will be mainly thought of as an end in itself. Later on we shall be able to compare predictions with facts, and so get down to the real business of decision-making.

Probability in diagram form

We now have two ways of estimating the probability connected with a particular experiment: experience and symmetry. The next stage is to apply this figure, however obtained, to situations more complex than the throw of a single coin or a single die. Initially, a graphical method is clearest. Consider first the case of a coin and a die thrown together. For the coin there are two results which are assumed equally likely, and for the die six. There are thus twelve possible results altogether, each of the two taken with each of the six in turn and, since conditions of symmetry still obtain (that is, there is no reason to favour one of the twelve rather than another), the probability of each is 1/12. This can be shown diagrammatically in the form below, which is called a **possibility space**. The dots represent the possible results of the experiment in co-ordinate form. If the probabilities were not all equal, a similar diagram could be drawn with numerical probabilities replacing the dots—this type is a **probability space**.

Figure 4.4   Possibility space for a coin and a die

A particular group of dots, representing one of the conditions in the problem, can be enclosed within a curve. The ones shown here are the results which include a head and those which include a six. So, to find the probability of throwing either a head or a six (or both), it is only necessary to count up the dots inside one or both curves and give the answer in twelfths—7/12.

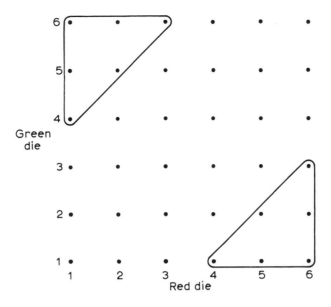

Figure 4.5   Possibility space for two dice

For a further illustration, consider a red die and a green die thrown together, shown in figure 4.5. From this, it is possible to read off the probabilities of throwing a double, a total of eight, at least one six, and similar combinations. All of these could easily be done without a diagram, but the real value of a possibility space will be seen if more complex problems are attempted without one. For example, find the probability that the difference between the scores is at least three (the problem illustrated in the diagram), or that the throw is a double or a score of nine.

The laws of probability

There are two special properties known as the addition and the multiplication laws of probability which look a trifle difficult when stated formally, but really are only applied common sense and hardly need a full proof. We have, in fact, used both of them already without saying so. In using a possibility space, we have added probabilities together to find the solution to a problem, with the dots representing the separate probabilities of the various forms of the solution. We could only do this if the sets of marked dots did not overlap—that is, in the coin-die problem we could not add the 6/12 for the coin directly to the 2/12 for the die, because this would have counted the six-and-head case twice. In other words, probabilities can be added if and only if they cannot occur together. Stated formally, the addition law of probabilities is as follows:

The probability that any one of two or more mutually exclusive events will occur is the sum of the probabilities of the events considered separately.

Thus, if a single die is thrown, it cannot show both a two and a three—the two possibilities exclude each other. Since the probabilities of each are 1/6, the probability of throwing either a two or a three is 1/3.

Also, when two independent events occur together, as with a six and a head, the probability turns out to be 1/12, the product of 1/6 and 1/2, illustrating the multiplication law of probabilities:

The probability that both (or all) of two or more independent events will occur is the product of the probabilities of the separate events.

The important word here is 'independent'. Two events are said to be 'statistically independent' if the probability of one is unaffected by the outcome of the other. If we throw a six first, this does not make a head any more or less likely; but if we are trying to draw aces from a pack, the probability of drawing an ace with the second card is affected by the result of the first draw.

Complementary probability

What is the probability of throwing at least one six if a die is thrown four times? 'At least one' in this context implies either 1, 2, 3 or 4, so if we approach the problem by the direct method, it is clearly a complicated one. But if a 'success' is to throw at least one six, a 'failure' is to throw none, a single possibility and one which is quite easy to work out. If p is the probability of a success, and we denote the probability of a failure by q, clearly $p+q = 1$. The probability of a non-six is 5/6 for one trial, and by the multiplication law $q = (5/6)^4$ for four independent trials. Hence $p = 1-(5/6)^4 = \dfrac{671}{1296}$. The value of q is called the **complementary probability** to p.

Conditional probability

This is the method used for the problem of successive non-independent trials, which was mentioned in the discussion of the multiplication law. The only things which need to be modified are the figures used for the second and subsequent events, because they are conditional on what has gone before—hence the name. Consider a bag containing six balls, three red and three black, identical except for colour, from which two balls are selected at random. What is the probability that both are black? For the first ball the probability is clearly $\frac{1}{2}$. If a success is still possible at this stage, then the bag must contain two black balls out of five, giving a probability of 2/5. The randomness of the choice itself is not affected, and so the multiplication law can now be used, giving an answer of $\frac{1}{2} \times 2/5$, or 1/5.

Next, what is the probability that four cards drawn randomly from a pack will be the four aces? The chance that the first is an ace is $\frac{4}{52}$; there are then 51 cards left, of which 3 are aces, so that the chance is reduced to $\frac{3}{51}$, and so on. The final answer is therefore $\frac{4 \times 3 \times 2 \times 1}{52 \times 51 \times 50 \times 49}$.

The last example in this section is one in which the best method is to find the complementary probability. It is assumed that the birthday of a person chosen at random is equally likely to be on each day of the year—this is not actually true, but it does not make very much difference; it is also assumed that a person born on February 29th is not eligible for selection. What is the probability that, in a random group of 30 individuals, at least two of them will have the same birthday? We find the probability that no two of them have the same birthday—in other words, they all have different birthdays. The first person can have any day. The second can have any of the remaining 364, for which the probability is $\frac{364}{365}$. The third can have any of the remaining 363, and so on until the last one has 336 to choose from. The probability that at least two will have the same birthday is therefore

$$1 - \frac{364 \times 363 \times 362 \dots 336}{365 \times 365 \times 365 \dots 365}$$

which is just over 0·7—a surprisingly high figure.

One point in problems involving conditional probability (and which can occur in other types also) sometimes gives difficulty. Consider the same balls-in-the-bag problem as before. Does it matter whether the balls are taken out together or one by one? In the original calculation, it was assumed that they were taken out one by one. Now suppose that they are taken out together, and consider the instant when the two balls are being held in the hand but have not yet been inspected. At this time, it may be argued that we have two random samples of one, and so the probability that each is black is $\frac{1}{2}$; since for a success both have to be black, the multiplication law gives an answer of $\frac{1}{4}$. Clearly this is wrong, because, on the same argument, the chance of a sample of four being all black is 1/16, even though there are only three black ones available! The fallacy is that the two are not *independently* random—the probability of either one of them being black is 2/5 if the other is black and 3/5 if it is red; but it has to be black if a success is to be possible. The previous answer therefore stands.

Combinations: the ^nC_r notation

We could have approached the problems of the last section from a different angle; consider, for example, the choosing of the four aces from a pack of cards. The sample of four can be drawn in a number of ways, all assumed equally likely; only one possible sample consists of the four aces, and hence we can write down the probability. The first card can be drawn in any one of 52 ways; having drawn that, the next can be drawn in 51 ways, and so on. The number of different samples is thus $52 \times 51 \times 50 \times 49$, and so the required probability should be the reciprocal of that. But it isn't—why not?

The fallacy lies in what is meant by a 'different' sample. Is the sample in which the aces are drawn in suit order club, heart, diamond, spade to be counted different from any other specified order? Clearly not, according to the requirements of the problem, but we have, in fact, done so. Now the number of different suit orders is $4 \times 3 \times 2 \times 1$, because the first one can be chosen in four ways, the next in three, and so on—the same line of reasoning as before. The total number of samples of four chosen *without respect to order* is therefore the previous figure divided by this one, which gives $\dfrac{52 \times 51 \times 50 \times 49}{4 \times 3 \times 2 \times 1}$, and the corresponding figure for the probability is the same as we derived before. The figure 1 is normally left in at this point, even though it does not affect the multiplication.

The number of ways of choosing a sample of r from n possibilities—

here 4 from 52—without respect to order is written ^nC_r, the letter C standing for combination. An alternative notation $\binom{n}{r}$ is sometimes preferred but is less economical in terms of typographical layout.

It will be seen that, when r is large, the form of ^nC_r given above is lengthy, and so a shorthand notation is used. This involves the factorial sign, $n!$ (or in manuscript $\lfloor n$), meaning the product of all the integers from n down to 1. Thus 6! is $6 \times 5 \times 4 \times 3 \times 2 \times 1 = 720$. The number of ways of choosing 4 from 52, given above, is then written as $\dfrac{52!}{4! \times 48!}$; if this is written out in full, all the figures from 48 down to 1 appear both top and bottom, so that, after division, only the numbers from 52 down to 49 are left on the top and the 4! on the bottom. Care in division with factorials is necessary: 8! divided into 9! goes 9 times, and 3! into 6! goes 120 times, not twice. For practical calculations, the tables of logarithms of factorials included as Table 6 at the back of the book will be found useful.

In the general case, the number of ways of choosing r items from n possibilities, without respect to order, is

$$^nC_r = \frac{n!}{r!\,(n-r)!}.$$

It follows immediately from this that $^nC_r = {}^nC_{n-r}$.

Probability using set notation

Readers familiar with the language of sets will have noticed the similarity between the possibility space and a Venn diagram. If $\mathscr{E}$ is a set from which a choice of one is to be made on an 'equally likely' basis, and if A and B are two subsets of $\mathscr{E}$, then the probability that the one chosen is a member of both A and B is $n(A \cap B)/n(\mathscr{E})$, and the probability that it is a member of either A or B (or both) is $n(A \cup B)/n(\mathscr{E})$. This gives a neat method of solving certain types of problems, such as dominoes, where the data is a little more complex than the ones met with earlier. There are 28 dominoes in a standard set, one with each possible combination of two scores from zero to six. If a domino is selected at random, what is the probability that one or both of its scores is a two or a three? In the Venn diagram, $\mathscr{E}$ is the set of dominoes, A is the set with a two and B is the set with a three. Since $n(A \cap B) = 1$, the three–two, and $n(A) = n(B) = 7$, the

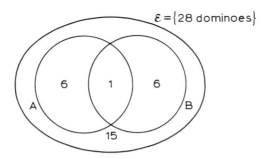

Figure 4.6   Venn diagram for the domino problem

numbers in the regions can be filled in at once, and the required probability is 13/28. You may like to try to do this with a possibility space for
comparison—it is a little longer, and not such a neat method. Those
who found it too easy might prefer to try the three-sided domino problem
at the end of the chapter, exercise 28.

Tree diagrams

One limitation of the possibility space is that, if there are three random
variables rather than two, a three-dimensional model is required. A
different type of diagram, however, can cover such cases, although it
imposes a different limitation instead—if the variables have more than
two possible values the diagram becomes rather complicated. It is called a
tree diagram on account of its shape, the 'tree' being drawn on its side or
upside down according to choice. In the following example we consider
the throw of three dice, a 5 or a 6 being counted a success, S ($p = \frac{1}{3}$),
and any other score a failure, F ($q = \frac{2}{3}$). Each die can give S or F, so the
possible results are SSS, SSF, SFS, SFF, FSS, FSF, FFS and FFF—
eight in all. The diagram looks like this:

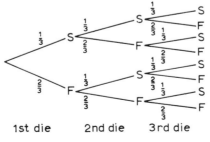

Figure 4.7   Tree diagram for three dice

The probabilities for any particular result may be found by following the lines through and multiplying together the probabilities along each line. Thus the probability of scoring two successes out of three is the sum of the probabilities of SSF, SFS and FSS, which gives 2/9. As a check, the probabilities of 0, 1, 2 and 3 successes (8/27, 4/9, 2/9 and 1/27) must total 1.

In this example, the experiment consists of three parts in which the probability is the same, but this need not be so. We could spin a coin, throw a die and cut a pack in turn, counting a success as say a head, a six and a spade respectively, and use a tree diagram to work out the results.

The 'Law of Averages'

This is an expression in common use and, if only for this reason, it is worth a mention. The great Dr Joad was once asked on a Brains Trust what it was, and gave the answer: 'The law of averages says that, if you spin a coin one hundred times, it will come down heads fifty times and tails fifty times.' He was, alas, quite wrong, as he must surely have known. A result of *exactly* 50-50 is in fact most unlikely, as the reader will discover later on in this book if he is not convinced already. The accuracy of the definition can be improved by adding the word 'about' before 'fifty' each time, but a much better answer can be given by using the idea of a limit mentioned earlier. Thus, if the Law of Averages means anything at all, it means that, if a coin is spun a very large number of times, the proportion of heads settles down to a steady value, which will be about $\frac{1}{2}$ if the coin is a fair one fairly spun.

Final note

The Laws of Probability which we have looked at are essentially statements of experience. So, also, is a statistical law which is well known to most laymen, though it must be admitted that the experience is of a somewhat subjective kind. It may conveniently be called the Law of Pure Cussedness, and it states that, if an experiment has two or more possible results of which one is much more inconvenient than the others, then the probability of the inconvenient one is much greater than $\frac{1}{2}$. In other words, a dropped piece of toast usually falls marmalade side down. Since the quantitative evidence in support of the Law is rather limited, it finds no further place in this book. But its properties are better understood by many laymen than is the case for other statistical laws, and some advanced exponents of it are able to turn it to their own advantage— for example, by carrying an umbrella to discourage rain.

Exercises

1 State what justification there is for stating a numerical value for the probability of the following; if there is no justification, say why not.
   (a) That an article coming off a production line is defective.
   (b) That the Olympic 1500 metres record will be broken at the next Games.
   (c) That a given die, known to be biased, will show a six.
   (d) That the standard rate of Income Tax will be altered in the next Budget.
   (e) That the 'Princes in the Tower' were murdered by Richard of Gloucester.
   (f) That the unborn child of a certain woman will be a boy.
   (g) That the total precipitation in Manchester next year will exceed 1·2 metres.

2 A die is thrown and a pack of cards is cut. Find the probabilities of (a) a six or a spade, or both; (b) a 3 or a 4 but not a spade.

3 A bag contains three red balls and seven black. Find the probability that a sample of four balls contains at least one red.

4 A set of dominoes has a number from zero up to nine (instead of the usual 6) on each half. Find the total number in the set, and the probability that a random domino will have at least one of either 8 or 9.

5 A committee of four is to be selected from a group of 7 men and 3 women. Find the number of ways in which the committee may be formed: (a) if there are no restrictions; (b) if there must be two men and two women; (c) if there must be at least one of each sex.

6 Two dice are thrown. Find the probability that (a) the scores are identical; (b) at least one of the scores is a six; (c) the total score is at least 10; (d) there is a difference of at least 2 between the scores; (e) the total score is a prime number.

7 A die is thrown three times. Find, by means of a tree diagram, the probabilities of 0, 1, 2 and 3 sixes—that is, the probability distribution for the number of sixes.

8 A bag contains three red balls and five black ones. Find the probability distribution for the number of reds in a sample of three.

9 A bag contains four red balls and eight black ones. Find the probability distribution for the number of reds in a sample of four.

10 A bag contains four red balls (score 3), four green (score 2), and four black (score 1). Find the probability distribution for the total score if a sample of three balls is drawn.

11 A bag contains 36 balls, six each of six different colours, and a sample of four is drawn. Prove that the probability that the balls are all different colours is over 200 times as great as the probability that they are all the same colour.

12 Four cards are drawn from a standard pack. Find the probabilities of the following: (a) four kings; (b) three kings and an ace; (c) two kings and two aces.

13 Thirteen cards, two of which are aces, are placed randomly (a) in a line, (b) in a circle. What is the probability that the two aces are adjacent in each case? What are the probabilities if there are n cards instead of 13?

14 A game is played as follows: a coin is spun until two tails have appeared, and the number of heads thrown counts as the score. Find the probability that the $(n+2)$'th coin is the second tail, and hence find the probability of a score of more than five.

15 Three balls are drawn together from a bag containing six black, three white and five red balls. Find the probability that two, and only two, are the same colour.   (JMB)

16 A box contains three red, four white and five blue balls, and a second box contains five red, six white and seven blue balls. One ball is drawn from each box: find the probability that they are the same colour.
                                                                          (JMB)

17 A man has five keys, only one of which will open a particular door. Find, for each integer k, the probability that the k'th key will open the door, if (a) the keys are tried at random; (b) the keys are tried in succession.   (JMB)

18 Consider the following experiments: (a) 12 cards are drawn at random, with replacement, from a standard pack of cards; (b) 4 cards are drawn at random, with replacement, from each of three packs; (c) 4 cards are drawn at random, without replacement, from each of three packs. In each case find the probability (to 3 d.p.) that the 12 cards are all different.   (JMB)

19 If A and B each throw a pair of unbiased dice, show that the probability that they obtain the same pair of numbers is approximately 0·05. Find also the probability that they obtain the same total with the two dice.   (JMB)

20 Two players compete by drawing, in turn and without replacement, one ball at random from a box containing four red and four white balls. The winner is the player who first draws a red ball. Calculate the probability that the winner is the player who makes the first draw. Find also the corresponding probability when the ball is replaced after each draw.   (JMB)

21 A, B and C throw a die in turn until a six appears, A throwing first. Find the probability that A will win on the first, second, third and n'th time round, and hence find each player's probability of winning the game.

22 Find the probabilities of the following five-card Poker hands, assuming that they are dealt directly from a shuffled pack; give answers in the form 1 in n, where n is taken to the nearest whole number. (a) A single pair, i.e. two of the same value and the other three all different; (b) two pairs; (c) three of a kind; (d) full house, i.e. a trio and a pair; (e) four of a kind; (f) flush, i.e. all cards of the same suit; (g) royal flush, i.e. AKQJ10 of any suit.

23 A 2p piece (diameter 2·6 cm) is rolled on to a large board divided into squares by two sets of parallel lines 4 cm apart. Find the probability that the coin comes to rest not touching a line.

24 A needle of length a is thrown randomly (both in position and direction) on to a surface ruled with lines distance D apart. ($D \geqslant a$.) Show that the probability that the acute angle between the needle and the lines is in the range θ to $\theta + \delta\theta$ is $2\delta\theta/\pi$, and find the probability that the needle crosses a line when falling at such an angle. Hence prove that the probability that the needle falls across a line is $2a/\pi D$. (This can be used to find a value for π, with considerable accuracy if done as a co-operative experiment; it is convenient to take $D = a$.)

25 Four people are to draw cards from a pack in succession, without replacement, ace counting high. The first player draws a ten; find the probability that he wins the draw, (a) without a recut; (b) with recuts as necessary, e.g. if one other player draws a ten, the two cut again. Give answers to 3 d.p.

26 A bag contains $3n$ balls, n red and $2n$ black, and a second bag contains $2n$ red and n black. A ball is transferred at random from the first bag to the second, and then a ball is transferred at random from the second back to the first. Show that the probability that a ball then drawn at random from the first bag is red is $(3n+2)/(9n+3)$.

27 A tennis match usually consists of either 3 or 5 sets, and ends when one side has won a majority of sets. If the probability of a side winning a set is a constant value p, show that the probability of a match going

to full length is $2pq$ in the case of a 3-set match and $6p^2q^2$ in the case of a 5-set match, where $q = 1-p$.   (OL)

28 Show that there are 84 triangular dominoes in a full set having 0 to 6 on each side; the order of the numbers is not taken into account. Find the probability that a random domino will (a) be a triple; (b) be a double; (c) include at least one six; (d) have 15 or more pips; (e) have each side a different score.

29 For a tetrahedral die (four faces, each an equilateral triangle), the 'score' is the face that is invisible. If four such dice are thrown, find the probability that (a) all scores are the same; (b) all scores are different; (c) there are exactly two different scores; (d) there are exactly three different scores. (If you find this too easy, try it for five or six tetrahedral dice, or for various numbers of ordinary dice.)

30 A game is played by throwing a die, with the following rules. If a player throws a 1, he scores a point and his turn ceases; if he throws a 2, he scores a point and has another throw; if he throws a 3, 4, 5 or 6 his turn ceases, and if he throws a 6 his score is also reduced to zero. Find the probability that a player will score (a) 1; (b) 2; (c) any number n; (d) 0.

31 In a game, a certain player wins a set with probability p when he has service and with probability $\frac{1}{2}p$ when he has not. A match consists of three sets if one player wins them all, or four sets if one player has then won three, or five sets otherwise. Service goes alternately. Find the probability that this player will win: (a) in three sets when he has first service; (b) in three sets when he has second service; (c) in four sets when he has first service; (d) in four sets when he has second service.

Chapter 5

The binomial distribution

In the section on tree diagrams in the last chapter, we considered the problem of three trials of an experiment for which $p = \frac{1}{3}$; we now go on to problems of the same type but rather greater complexity. As the number of trials increases, the tree diagram soon becomes too complicated to use, and the first step towards a new method is to look for a pattern in the probabilities. Using p and q for the probabilities of success and failure respectively, rather than numerical values, we can draw up a table like this:

Number of successes:	0	1	2	3	4
Probability in 1 trial:	q	p			
Probability in 2 trials:	q^2	$2qp$	p^2		
Probability in 3 trials:	q^3	$3q^2p$	$3qp^2$	p^3	
Probability in 4 trials:	q^4	$4q^3p$	$6q^2p^2$	$4qp^3$	p^4

You should be able to check the first three lines without difficulty using a tree diagram similar to figure 4.7. Now consider the fourth line, and in particular the probability of getting exactly 2 successes out of 4, which can be worked out from the previous line as follows. To achieve 2 in 4 it is necessary to get either 1 success in the first 3, followed by a success (probability $3q^2p \times p = 3q^2p^2$), or 2 successes in the first 3 followed by a failure ($3qp^2 \times q = 3q^2p^2$). The sum of these two is $6q^2p^2$ as shown.

We can now give the general way of getting each line from the one above. First write down the q's and p's, following the obvious pattern. Then work out the coefficients in turn, by adding together the coefficient immediately above and the one to its left. A blank counts as zero, of course. Thus the coefficients for the next line will be 1, 5, 10, 10, 5, 1. A table of these coefficients appears as Table 2 at the end of the book, and many readers will recognise them as the numbers in Pascal's Triangle.

Another pattern can also be seen in the table. The lines are, in turn, the full forms of $(q+p)$, $(q+p)^2$, $(q+p)^3$ and so on. No rigorous proof will be given at this point, but another way of approaching the subject will be discussed later.

We now have a probability distribution which is called the **binomial distribution**, which can be expressed as follows:

In n trials of an experiment in which the probability of a success is p, the probabilities of 0, 1, 2 ... n successes are the terms of the binomial expansion of $(q+p)^n$, where $q = 1-p$.

Example 1 If a pack of cards is cut 5 times, being shuffled between each cut, what are the probabilities of cutting 0, 1, 2, 3, 4 and 5 spades—that is, the probability distribution for the number of spades? Here $p = \frac{1}{4}$ and $q = \frac{3}{4}$, and we require the expansion of $(\frac{3}{4}+\frac{1}{4})^5$. Since the denominator is the same for both terms, it is convenient to write the expression as

$$(\tfrac{1}{4})^5(3+1)^5 = (243+5\times81+10\times27+10\times9+5\times3+1)/1024,$$

so that the respective probabilities are $\dfrac{243}{1024}$, $\dfrac{405}{1024}$, $\dfrac{270}{1024}$, $\dfrac{90}{1024}$, $\dfrac{15}{1024}$,

and $\dfrac{1}{1024}$. As a check, the sum of these is 1.

Example 2 If six dice are thrown, what is the probability of getting two or more sixes? Here the complementary probability, that of getting none or one, is easier to find—it is

$$\left(\frac{5}{6}\right)^6 + 6\left(\frac{5}{6}\right)^5\left(\frac{1}{6}\right) = \frac{15\,625+18\,750}{46\,656} = \frac{34\,375}{46\,656}.$$

The answer is therefore

$$\frac{12\,281}{46\,656} = 0\cdot263.$$

Example 3 If 52% of live births are male, what is the probability that, in a family of four children with no twins, there will be (a) exactly 1 boy and (b) exactly 2 boys? Since $p = 0\cdot52$, the answers are (a) $4\times0\cdot48^3\times0\cdot52 = 0\cdot23$, and (b) $6\times0\cdot48^2\times0\cdot52^2 = 0\cdot37$.

Example 4 In a manufacturing process, 1% of articles are defective. What is the probability that, in a batch of 50, there will be more than one defective? The probability is

$$1-(0\cdot99^{50}+50\times0\cdot01\times0\cdot99^{49}) = 1-1\cdot49\times0\cdot99^{49}.$$

The term 0.99^{49} can be worked out by logarithms, noting that the log of 0.99, $\bar{1}.9956$, is the same as -0.0044. Multiplication by 49 gives -0.2156, i.e. $\bar{1}.7844$. Hence the answer is 0.09. This is not an accurate answer, because the log multiplied by 49 is correct only to two figures; taking the two limits -0.00435 and -0.00445, the answer is in the range 0.0866 to 0.0965—quite an instructive example for those who are confident that four-figure tables always give three-figure accuracy.

Alternative derivation of the binomial distribution

Again suppose that we have n trials, with probabilities p and q as before, and consider the probability of getting exactly x successes, and thus $n-x$ failures. The probability that the *first* $n-x$ trials are all failures and the rest successes is $q^{n-x}p^x$, by the multiplication law. But this is only one of the possible ways; the total is the number of ways of arranging x successes in the n trials without respect to order, which is ^nC_x. The probability is therefore

$$^nC_x q^{n-x} p^x = \frac{n!}{x!\,(n-x)!} q^{n-x} p^x.$$

Those who have met the binomial theorem before will recognise this as the general term of $(q+p)^n$. Those who have not should practise using it as a formula for working out the binomial coefficients for a few values of n, say up to 12 or so, using Table 2 to check their answers.

Example 5 What is the probability that a family of 9 children will have exactly 5 boys, if the probability that each one is a boy is 0.52? This is $^9C_5 \times 0.48^4 \times 0.52^5$, which is about 0.25. One point of difficulty in examples of this type is to get the p and the q the right way round, but a little ingenuity avoids even the need to decide which is which. $^9C_5$ is the same as $^9C_4$, so that either can be used; 0.48 is the probability of a girl, and there are 4 girls, so write 0.48^4; and similarly for the boys.

Example 6 This involves three possible outcomes from each experiment. Six dice each have three faces coloured black, two green and one red. What is the probability that, when the six are thrown together, the top faces will be two of each colour?

We work in two stages. First find the probability that two are black and four are non-black, which is $^6C_2(\frac{1}{2})^3(\frac{1}{2})^3 = \dfrac{15}{64}$. Then exactly two of the four have to be green, for which the probability is now $\frac{2}{3}$, giving $^4C_2(\frac{2}{3})^2(\frac{1}{3})^2 = \frac{8}{27}$. The answer is then $\frac{15}{64} \times \frac{8}{27}$, which is $\frac{5}{72}$. A shorter method (if you can see it) gives $\frac{6!}{2!2!2!}(\frac{1}{2})^2(\frac{1}{3})^2(\frac{1}{6})^2$ directly.

Expectation

If we imagine an experiment in which a fair coin is thrown 100 times, then it would be quite usual to say that we would *expect* to get about 50 heads. To be more precise, we could say that the average number of heads in a long run of such experiments might be *expected* to be very close to 50. Or to be more precise still, we could say that the mean value of the probability distribution of the number of heads is 50. It is often useful to have a single term meaning 'the mean value of the probability distribution' of a given random variable, and for this, the word **expectation** is used. It is similar in meaning to the word 'expect', as used in the two sentences above. The notation $E(x)$ is often used for the expectation of a random variable x.

As an example, consider a gambler who draws a ball at random from a bag containing one black ball and nine red, who is to receive £1 if he draws the black. His expectation is the amount he can expect to receive per game, on average, over a long series of games. This is clearly £0·10, since he can expect to win once in every ten games, taken overall. It will be seen that the expectation is also the fair stake in the game—that is, the stake which allows no long-term advantage to either side. In this case the random variable (the winnings) takes value £1 with probability 0·1, and value zero otherwise, so that the expectation is 0·1 × £1 = £0·10, the mean value of the random variable.

(Those who have ventured into the hazardous business of gambling will know that, in practice, any stake is rather more than the expectation. Bookmakers don't take chances, whatever their clients may do.)

Two other similar terms are also used. First, **expected value** is a synonym for expectation which is sometimes convenient, but can be misleading in that it appears to imply that the actual value given is in some way expected to occur. This is not so, and it may not even be physically possible for it to do so—for example, the expectation of the number of heads in five coins is $2\frac{1}{2}$.

Second, for a number of trials of an experiment whose result has a given probability distribution, we can give the **expected frequency** of each type of result. If five coins are thrown 96 times, the expected frequency of the five-head result is $96 \times 1/32 = 3$. Where figures do not turn out to be whole numbers, they are often suitably rounded off. The term 'theoretical distribution' is also used in this context, but has little to commend it.

Example 7 On the throw of a die, £5 is to be paid for a six, £4 for a five and £3 for a four. What is the expectation? The chance of each is 1/6, so the expectation is $£(5/6 + 4/6 + 3/6) = £2$. (This illustrates, incidentally, the way in which the expectations for the different possible outcomes of an experiment may be added.)

Example 8 A standard pack of cards is cut five times. What is the expectation of the number of court cards? The probability is 3/13 for each cut, and so the expectation is $5 \times 3/13 = 1\frac{2}{13}$ court cards.

Example 9 Six coins are thrown 200 times. Find the expected frequency distribution. The probabilities are the terms of the expansion of $(\frac{1}{2} + \frac{1}{2})^6$, which are 1/64, 6/64, 15/64, 20/64, 15/64, 6/64 and 1/64 respectively. These are multiplied by 200 to give the expected frequency distribution, that is, $3\frac{1}{8}$, $18\frac{3}{4}$, $46\frac{7}{8}$, $62\frac{1}{2}$, $46\frac{7}{8}$, $18\frac{3}{4}$, and $3\frac{1}{8}$. Rounded off to whole numbers, the answer is thus 3, 19, 47, 62, 47, 19 and 3, the $62\frac{1}{2}$ being taken as 62 rather than 63 in order to make the total 200. (This distribution can easily be compared with an observed distribution, and later on we shall use a method for making such comparisons quantitatively.)

Example 10 A pack is to be cut five times, and £20 is to be paid for every spade. What is the expectation?

We shall use two methods for this. First, consider the separate expectations for the six possible different results, for 0 to 5 spades respectively. The probabilities have already been worked out in example 1, and there is a probability of 405/1024 of winning £20, 270/1024 of winning £40, and so on. The expectation is therefore

$$£(405 \times 20 + 270 \times 40 + 90 \times 60 + 15 \times 80 + 1 \times 100)/1024,$$

which is £25.

But it is much easier to consider the expectations of the separate *cuts*. The probability is $\frac{1}{4}$ for each cut, giving an expectation of £5 per cut, a total of £25 for five cuts. This indicates a short way of finding the **mean or** expectation of a binomial distribution, and we turn to this next.

The mean of a binomial distribution

In the example just quoted, the expectation is £5 per trial when the return is £20 for a success. Another way of expressing the same fact is to say that the expectation of the *number* of successes is $\frac{1}{4}$ per trial or $1\frac{1}{4}$ in five trials. We now put this result in more general terms: in n trials of an experiment in which the probability of a success is p, the expectation of the number of successes is np. Since the expectation is the same thing as the mean of a probability distribution, the mean number of successes for a binomial distribution is np.

Working out the SD of a binomial distribution is not so easy and, to show the method for this, we shall first work out the mean in a new way, both for the special case $n = 4$ and the general case. It is convenient to write the distribution in the form $(q+pt)^n$ so that, when multiplied out, the coefficients of t^0 (i.e. the constant term), t, t^2 ... are the probabilities of 0, 1, 2 ... successes. This is called a **probability generating function** (p.g.f.).

The coefficients are shown for brevity as f_0, f_1, f_2 .... So for $n = 4$ we have the identity

$$(q+pt)^4 = f_0 + f_1 t + f_2 t^2 + f_3 t^3 + f_4 t^4.$$

The expectation of the number of successes is worked out as follows. There is a probability f_0 of no successes, giving, of course, no expectation; the probability of one success is f_1, giving an expectation f_1; that of two successes is f_2, the expectation being $2f_2$, and so on. This gives the series $f_1 + 2f_2 + 3f_3 + \ldots + nf_n$, adding the expectations as in example 7, or, more briefly, Σfx. The same result follows if we consider the usual form for a mean, $\Sigma fx/\Sigma f$ since, for a probability distribution, $\Sigma f = 1$.

The expression can be obtained most easily by differentiating the identity above with respect to t. (The differentiation is merely a device for producing the algebraic result we want, and has got nothing to do with the probabilities themselves.) This gives:

$$4p(q+pt)^3 = f_1 + 2f_2 t + 3f_3 t^2 + 4f_4 t^3.$$

Now, putting $t = 1$ (which makes the p.g.f. into a probability distribution once again) and using the fact that $p+q = 1$, the result is $4p = \Sigma fx$; hence the expectation or mean is $4p$.

In general,

$$(q+pt)^n = \Sigma f t^x$$

$$\therefore np(q+pt)^{n-1} = \Sigma fx t^{x-1}$$

$$\therefore \mu = \Sigma fx = np. \tag{5.1}$$

Greek letters are normally used for the constants of a probability distribution, and so μ (mu) is used here for the mean; similarly σ (the small sigma) for the SD in the next section.

The standard deviation of a binomial distribution

The starting point is equation 3.5, slightly adapted by using the facts that $n = \Sigma f = 1$ and that $\bar{x}$ is here μ, giving $\sigma^2 = \Sigma f x^2 - \mu^2$. We shall find $\Sigma f x^2$, first with $n = 4$ as before and then in the general case. First we multiply by t the line obtained by differentiation in the proof above, which gives

$$4pt(q+pt)^3 = f_1 t + 2f_2 t^2 + 3f_3 t^3 + 4f_4 t^4.$$

This is differentiated once again, and then t put equal to 1.

$$4p(q+pt)^3 + 12p^2 t(q+pt)^2 = f_1 + 2^2 f_2 t + 3^2 f_3 t^2 + 4^2 f_4 t^3$$

$$\therefore 4p + 12p^2 = \Sigma f x^2.$$

$$\therefore \sigma^2 = \Sigma f x^2 - \mu^2 = 4p + 12p^2 - 16p^2 = 4p(1-p) = 4pq.$$

Finally repeat for the general case.

$$npt(q+pt)^{n-1} = \Sigma f x t^x$$

$$\therefore np(q+pt)^{n-1} + n(n-1)p^2 t(q+pt)^{n-2} = \Sigma f x^2 t^{x-1}$$

$$\therefore np + n(n-1)p^2 = \Sigma f x^2$$

$$\therefore \sigma^2 = \Sigma f x^2 - \mu^2 = np + n^2 p^2 - np^2 - n^2 p^2 = np(1-p)$$

$$\therefore \sigma^2 = npq. \tag{5.2}$$

Thus the SD of the binomial distribution equals $\sqrt{(npq)}$. Many readers will not require the proof, but the result should be known.

We cannot make very much use of it at the moment, but later on it will enable us to carry out tests of significance—for example deciding whether the results of a coin-spinning experiment justify a conclusion that the coin is biased.

Example 11 Find the mean and SD of the probability distribution for a die thrown 180 times, a six being counted as a success. Here $p = \frac{1}{6}$ and $n = 180$, so that the mean is $np = 30$ and the SD $= \sqrt{(npq)} = \sqrt{(180 \times \frac{5}{6} \times \frac{1}{6})} = 5$. To give a rough idea of what this means, we can use a fact which was mentioned without proof in chapter 3: that, in many distributions (including this one), about $\frac{2}{3}$ of the distribution lies within ± 1 SD of the mean. So, if a large number of people all throw an unbiased die 180 times, about two-thirds of them may be expected to get a number of sixes in the range 25 to 35. (A small correction for continuity is ignored here.)

The symmetrical binomial distribution

If $p = \frac{1}{2}$, the probability of x successes is the same as the probability of x failures, and the distribution is therefore symmetrical about the mean of $n/2$. The general term of the distribution is $^nC_x(\frac{1}{2})^n$ and so, for a given value of n, the form of the distribution can be found by drawing a graph of ^nC_x against x. A sketch for $n = 25$ is shown in figure 5.1, using the data of Table 2.

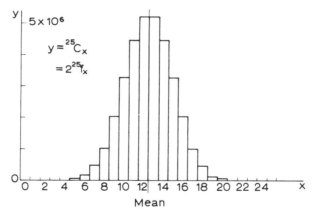

Figure 5.1   Binomial distribution for $n = 25$ and $p = \frac{1}{2}$

The mean of $12\frac{1}{2}$ is shown at the boundary between 12 and 13. The x-scale may be considered as continuous, with the block for, say, 12 covering everything from $11\frac{1}{2}$ to $12\frac{1}{2}$, following the method outlined in chapter 4. Thus the region between the ± 1 SD points includes half the area of the 10 and 15 blocks (the SD being $2\frac{1}{2}$) and all the blocks in between. The proportion of the distribution between these limits is therefore

$$\frac{1}{2^{25}}(2 \times 5200\,300 + 2 \times 4457\,400 + 3268\,760) = \frac{705\,755}{1048\,576} = 0.673,$$

which is very close to the figure $\frac{2}{3}$ already suggested.

It is instructive to compare this with the distribution for $n = 16$ and $p = \frac{1}{2}$, for which the SD is 2. It will be seen from a quick sketch that the form of it is very much the same—a bell-shape, with the tails on each side becoming too small to draw accurately more than 3 SD's away from the mean. The best way to compare the two is to use different scales for x so that the same length represents 1 SD on both—say 2 cm per unit for $n = 25$ and 2·5 cm per unit for $n = 16$, giving 5 cm for each SD. The vertical scales must, of course, be different as well—to be precise, the ratio between them should be $2^9 \times \frac{4}{5} = 409\cdot6$. Slide rule accuracy is sufficient.

You will find that the shapes are very similar and, if a curve is drawn through the mid-points of the tops of the blocks (this is not justifiable theoretically, but never mind), the two are barely distinguishable. This suggests that, when n is very large indeed, then, for $p = \frac{1}{2}$, the probability distribution tends towards a limit. The limit is, in fact, a curve known as the normal distribution, which is the subject of chapter 7.

For large values of n and values of p not too far away from $\frac{1}{2}$, the central section will still be reasonably symmetrical, though the tails, of course, are not. For example, the figures below give the expected frequency distribution for the 1 048 576 different ways of cutting a pack for say a spade 10 times, with $p = \frac{1}{4}$. Starting with no successes, the frequencies are 59 049, 196 830, 295 245, 262 440, 153 090, 61 236, 17 010, 3240, 405, 30 and 1. A graph of these is reasonably close to symmetry in the range $x = 0$ to $x = 5$. In the limit, as n again becomes very large, and even though p does not equal $\frac{1}{2}$, the curve is once again symmetrical.

A note on the different kinds of distribution

We have so far met distributions of two distinct types, each in two or three forms, and so it seems useful to give a summary of them.

First, we have the *observed* frequency distribution, which usually takes a form such as that of figure 1.1, and consists of experimental data arranged in groups according to size, the frequency being the number of observations in each group. Alternatively, we can arrange the data in the form of a frequency density distribution, in which the proportion of the total is shown for each group, as in figure 1.3.

The second type, which must be clearly distinguished from an observed distribution, is the *probability* distribution, which may be either discrete or continuous. Derived from this, we have the expected frequency distribution. The probability distributions which we have met so far have been useful for providing a model which may be compared with an observed distribution, and this will also be true for the probability distributions we shall meet in the next two chapters. After these, we shall consider a probability distribution of a different, and most important, kind—this is the sampling distribution, which forms the basis of the greater part of the rest of the book.

Exercises

1 A die is thrown four times, and a 5 or 6 is counted a success. Find the probability distribution for the number of successes.

2 A pack is cut five times. Find the probability distribution for the number of spades.

3 Five dice are thrown. Find the probability of (a) no sixes, and of (b) at least two sixes.

4 Ten fair coins are thrown. Find the probability of eight or more heads.

5 A pack is cut eight times. Find the probability of at least two aces.

6 In a certain experiment, the probability of a success is 2/7, and the experiment is repeated 12 times. Find the most probable number of successes, and the probability of getting that number.

7 In an experiment the probability of a success is $\frac{3}{4}$. Show that in 15 trials there are two numbers of successes which are equally the most likely, and find the value of the probability of each.

8 Find the probability of throwing exactly n heads if a coin is spun $2n$ times, and show that this is a decreasing function of n. Show also that the probability of getting exactly 50 heads in 100 throws is about 0·08.

9 Taken over a long run, the proportion of defectives in a manufacturing process is 5%. In a random sample of 25, find the probability of (a) no defectives, (b) exactly two defectives, and (c) more than two defectives.

10 A counter is placed at the origin on a Cartesian graph, and a coin is spun to determine its movement: for a head it moves one unit in the positive x-direction and for a tail it moves one unit in the positive y-direction. Find the probability that, after eight throws, the counter is more than six units from the origin in a direct line.

11 Find the probabilities (to 3 d.p.) of obtaining (a) at least one six in 6 throws of a die; (b) at least two sixes in 12 throws; (c) at least three sixes in 18 throws; (d) at least one double-six in 36 throws of two dice.

12 An experiment can have one of three results, A, B and C, with probabilities $\frac{1}{2}$, $\frac{1}{3}$ and $\frac{1}{6}$ respectively. Find the probability of getting 3 A's, 2 B's and a C in six trials.

13 Articles are manufactured in large batches, and it is desired to keep the proportion of defective ones below 5%. A sample is drawn and tested, and if it contains any defectives the whole batch is tested. How many should be in the sample in order to give a 95% probability of detecting a defective rate of 5%?

14 With the same conditions as in question 13, how many should be in the sample in order to give a 90% probability of detecting a defective rate of 1%?

15 A gun is engaging a target, and it is required to make at least two direct hits. It is estimated that the probability of a direct hit with a single round is constant and equal to $\frac{1}{4}$. A burst of five rounds is fired, and if at least two direct hits are scored firing ceases; otherwise a second burst of five rounds is fired. Find the chance that at least two direct hits will be scored, (a) only five rounds being fired, (b) ten rounds having to be fired.   (O & C)

16 In an experiment, the probability of a success is 0·1. How many trials are necessary in order to have a 99% probability of at least one success?

17 In an experiment, the probability of a success is $\frac{1}{4}$. How many trials are necessary in order to have a 99% probability of at least two successes?

18 A coach holds 48 passengers. On average, one passenger in every 16 who books a seat fails to turn up. Stating carefully any assumptions made, find the probability that, if the company accepts 50 bookings, there will not be enough seats.

Chapter 6

The Poisson distribution

Consider the following problems:

(a) Twelve fair dice are thrown; find the probability distribution for the number of sixes. This is a binomial distribution for which $p = 1/6$, given by the coefficients of t in the expansion of $(5/6 + t/6)^{12}$.

(b) In a certain type of mass-produced article, 1% of the output is defective; find the probability distribution for the number of defectives in a sample of 200. By the same method, the distribution is the expansion of $(0\cdot99 + 0\cdot01t)^{200}$.

(c) In a large book of mathematical tables, there are about 4000 digits per page, and the average number of errors per page is 2; find the probability distribution for the number of errors per page. This is a binomial distribution again, with $p = 0\cdot0005$, given by the expansion of $(0\cdot9995 + 0\cdot0005t)^{4000}$.

These examples are chosen because they all have a mean of 2. The table below shows the comparative probabilities as far as they are measurable to three decimal places.

	f_0	f_1	f_2	f_3	f_4	f_5	f_6	f_7	f_8
(a)	0·112	0·269	0·296	0·197	0·089	0·028	0·007	0·001	0·000
(b)	0·134	0·270	0·271	0·181	0·090	0·036	0·012	0·003	0·001
(c)	0·135	0·271	0·271	0·180	0·090	0·036	0·012	0·003	0·001

The similarity between (b) and (c) suggests that the binomial distribution approaches a limit as the probability becomes very small, if the average number of successes np is constant. The resulting distribution was discovered by S. D. Poisson and is named after him. The derivation of it is not particularly easy—readers who are happy to accept the results without proof should note the formulae numbered 6.1 and 6.2, and otherwise skip to the next section.

The starting point is a binomial distribution in which p is very small and the mean np is fixed; the latter will be labelled μ, following the convention of using a Greek letter for the constants connected with a probability distribution. The general term of the binomial is

$$\frac{n!}{x!\,(n-x)!}q^{n-x}p^{x}$$

$$= \left(\frac{1}{x!}\right)(n)(n-1)(n-2)(n-3)\ldots(n-x+1)\left(\frac{p}{1-p}\right)^{x}(1-p)^{n}$$

writing $(1-p)$ for q. Since there are x terms from the n to the $(n-x+1)$ inclusive, these terms can be paired off each with one of the type of the next term, giving

$$\left(\frac{1}{x!}\right)\left(\frac{np}{1-p}\right)\left(\frac{np-p}{1-p}\right)\left(\frac{np-2p}{1-p}\right)\ldots\left(\frac{np-xp+p}{1-p}\right)(1-p)^{n}$$

Now write np as μ, and let p tend to zero, and the term becomes

$$\left(\frac{1}{x!}\right)\mu^{x}\,\underset{p\to0}{\text{Lt}}\left\{(1-p)^{\mu/p}\right\}.$$

The limit of the last part is $e^{-\mu}$, the negative exponential function; a proof of this can be found in any advanced algebra text book. The distribution is thus as follows, taking x as $0, 1, 2 \ldots$:

$$e^{-\mu}\left\{1+\mu+\frac{\mu^{2}}{2!}+\frac{\mu^{3}}{3!}+\ldots+\frac{\mu^{x}}{x!}+\ldots\right\} \tag{6.1}$$

It is of course an infinite one, but in practice the terms soon become too small to be worth calculation unless the mean is more than 8 or so. An alternative method for large values of μ, using the normal distribution, will be described in the next chapter.

The variance follows at once from the formula derived for the binomial:

$$\sigma^{2} = \underset{p\to0}{\text{Lt}}\left(npq\right) = \underset{p\to0}{\text{Lt}}\left\{\mu(1-p)\right\} = \mu, \tag{6.2}$$

i.e. the variance is numerically equal to the mean.

An alternative derivation is by means of the p.g.f. for the binomial, given in chapter 5. Again, we consider the limit $p \to 0$, with $np = \mu$ so that $n \to \infty$. The p.g.f. for the Poisson is

$$\underset{p\to0}{\text{Lt}}\,(q+pt)^{n} = \underset{p\to0}{\text{Lt}}\left\{q^{n}\left(1+\frac{pt}{q}\right)^{n}\right\} = \underset{p\to0}{\text{Lt}}\left\{(1-p)^{\mu/p}\left(1+\frac{pt}{q}\right)^{\frac{q}{pt}\times\frac{\mu t}{q}}\right\}$$

$$= e^{-\mu}e^{\mu t} = e^{-\mu}\left\{1+\mu t+\frac{\mu^{2}t^{2}}{2!}+\frac{\mu^{3}t^{3}}{3!}+\ldots\frac{\mu^{x}t^{x}}{x!}+\ldots\right\}$$

which gives equation 6.1 when t is put equal to 1.

The mean and variance can be derived from the series directly, both by differentiation of the p.g.f. as for the binomial, and by the following method.

$$\text{Mean} = \Sigma fx = e^{-\mu}\left\{0 \times 1 + 1 \times \mu + 2\frac{\mu^2}{2!} + 3\frac{\mu^3}{3!} + \ldots + \frac{\mu^x}{x!} + \ldots\right\}$$

$$= e^{-\mu}\left\{1 + \mu + \frac{\mu^2}{2!} + \ldots + \frac{\mu^{x-1}}{(x-1)!} + \ldots\right\}\mu$$

$$= e^{-\mu}e^{\mu}\mu = \mu$$

$$\text{Variance} \quad \sigma^2 = \Sigma fx^2 - \bar{x}^2 = \Sigma fx^2 - \mu^2$$

$$= e^{-\mu}\left\{0^2 \times 1 + 1^2 \times \mu + 2^2\frac{\mu^2}{2!} + 3^2\frac{\mu^3}{3!} + \ldots + x^2\frac{\mu^x}{x!} + \ldots\right\} - \mu^2$$

$$= \mu e^{-\mu}\left\{1 + 2\mu + 3\frac{\mu^2}{2!} + \ldots + x\frac{\mu^{x-1}}{(x-1)!} + \ldots\right\} - \mu^2$$

$$= \mu e^{-\mu}\left\{1 + \mu + \frac{\mu^2}{2!} + \ldots + \frac{\mu^{x-1}}{(x-1)!} + \ldots\right.$$

$$\left. + \mu + 2\frac{\mu^2}{2!} + \ldots + (x-1)\frac{\mu^{x-1}}{(x-1)!} + \ldots\right\} - \mu^2$$

The second part of the long expression in this bracket has already appeared in the calculation of the mean, and so

$$\sigma^2 = \mu e^{-\mu}\{e^{\mu} + \mu e^{\mu}\} - \mu^2 = \mu \quad \text{as before.}$$

The Poisson distribution in practice

Example 1 Consider the two-errors-per-page problem of the previous section. Since $\mu = 2$, we get the series

$$e^{-2}\left\{1 + 2 + \frac{2^2}{2!} + \frac{2^3}{3!} + \frac{2^4}{4!} + \ldots\right\}$$

From Table 11, $e^{-2} = 0.1353$. The easiest method of calculation is to use each term to find the next one, by multiplying successively by 2, $\frac{2}{2}$, $\frac{2}{3}$, $\frac{2}{4}$ and so on. When the mean is not a convenient round number, the process is not quite so easy, but it remains the best method. If the calculation is continued until the terms become very small, a check is possible by comparing the sum with 1. The reader is left to work this distribution out himself; rounded off to three places of decimals, the figures are the same as those obtained by the binomial.

Example 2 Seismograph records over a number of years indicate that, in a particular place, the average number of earthquakes of over a given strength is 1·3 per year. What is the probability that in the next twelve months there will be more than three earthquakes? (Note that this is genuinely the limiting case of a binomial and not an approximation: '*p*' is the probability of an earthquake occurring at a given instant in time, which is infinitesimally small.)

$$f_0 = e^{-1\cdot3} \qquad = 0\cdot2725$$

$$f_1 = 0\cdot2725 \times 1\cdot3 = 0\cdot3542$$

$$f_2 = 0\cdot3542 \times \frac{1\cdot3}{2} = 0\cdot2303$$

$$f_3 = 0\cdot2303 \times \frac{1\cdot3}{3} = \underline{0\cdot0998}$$

$$\overline{0\cdot9568}$$

Hence the answer is 0·043—about once in every 23 years.

The Poisson expected frequency distribution

The figures we have worked out so far have all been probabilities. We can use them to predict what would happen in practice, in much the same way that the value $p = \frac{1}{2}$ for a head in a coin-spinning experiment implies about 50 ($= \frac{1}{2} \times 100$) heads in 100 trials. Usually some rounding off is required. The Poisson probabilities, multiplied in turn by the number of trials, thus give the corresponding expected frequency distribution. Differences between 'theory' and 'practice'—usually a misleading contrast anyhow—are to be expected, and later on we shall be able to test whether they are large enough to matter.

The classic example which follows has appeared in almost every book dealing with the Poisson, but is too good to leave out. In the Prussian army, records were kept of deaths due to kicks of a horse, for ten corps over 20 years, and the 200 observations were divided up according to the number of deaths per corps per year as follows:

Number of deaths:	0	1	2	3	4	5 or more
Number of observations:	109	65	22	3	1	Nil

Since the probability of a given soldier being killed is small, then, assuming that the corps were of more or less uniform size and type, the figures might be expected to follow a Poisson distribution. We therefore compare it with the expected frequency distribution which has the same mean. The total number of deaths was 122, giving a mean of 0·61.

$$f_0 = e^{-0.61} \qquad\quad = 0.5434 \qquad\qquad \therefore 200f_0 = 109$$

$$f_1 = 0.5434 \times 0.61 = 0.3315 \qquad\qquad \therefore 200f_1 = 66$$

$$f_2 = 0.3315 \times 0.61 = 0.1011 \qquad\qquad \therefore 200f_2 = 20$$

$$f_3 = 0.1011 \times 0.61 = 0.0206 \qquad\qquad \therefore 200f_3 = 4$$

$$f_4 = 0.0206 \times 0.61 = 0.0033 \qquad\qquad \therefore 200f_4 = 1$$

Agreement with the actual values is remarkably close.

When the Poisson distribution is used

The primary use of the Poisson distribution is where a series of events occurs at independent random times, such as earthquakes, telephone calls through an exchange, radioactive emission, or vehicles passing a given point if traffic is light. The distribution may also be used as an approximation to the binomial if the probability is 0·05 or less, typically in sampling for defective items in manufacturing, when the rejection rate is usually less than 5%. An indication that the Poisson is worth investigation may be given by the shape of the observed distribution, which takes one of the forms shown in figure 6.1.

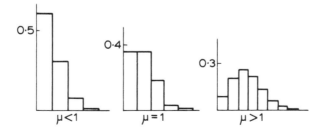

Figure 6.1   Typical forms of the Poisson distribution

The sum of two or more Poisson variates

Suppose a large batch of manufactured items has to be checked for defectives in two separate and independent ways—say length and width. If the mean numbers of defectives in a batch of 1000 are 2 for length and 3 for width, we can determine the probability of obtaining a given number of defectives overall: 4, for example. This is the sum of the probabilities of getting 4 out of tolerance for length and none for width, 3 for length and 1 for width, and so on. Since the probability for an individual item is small in both cases, all the probabilities are terms of a Poisson distribution, and the total is

$$e^{-2}\frac{2^4}{4!}e^{-3} + e^{-2}\frac{2^3}{3!}e^{-3}\frac{3}{1!} + e^{-2}\frac{2^2}{2!}e^{-3}\frac{3^2}{2!} + e^{-2}\frac{2}{1!}e^{-3}\frac{3^3}{3!} + e^{-2}e^{-3}\frac{3^4}{4!}$$

$$= e^{-2}e^{-3}\left(\frac{1}{4!}\right)\left\{2^4 + \frac{4!}{3!1!}2^3 \times 3 + \frac{4!}{2!2!}2^2 \times 3^2 + \frac{4!}{1!3!}2 \times 3^3 + 3^4\right\}$$

$$= e^{-5}\frac{(2+3)^4}{4!} = e^{-5}\frac{5^4}{4!}.$$

This quantity is exactly the probability of four defectives if the distribution is a simple Poisson with its mean the sum of the means of the two separate parts, i.e. 5. A general proof, on the lines of this numerical illustration, can be worked out without serious difficulty, and it can be extended at once to cover more than two constituent distributions. We have, then, the general result that the sum of two or more Poisson variates is itself a Poisson variate with mean equal to the sum of the means of the constituent variates.

This result enables the Poisson distribution to be used (as we have, in fact, tacitly assumed) in cases where the probabilities of the individual separate events are not the same, even though they are all very small. In the Prussian Army, to use the example quoted earlier on, some horses were doubtless more dangerous than others, and some soldiers more skilful than others at keeping out of their way; but a Poisson distribution for the numbers of deaths by horse-kicks was to be expected all the same.

Exercises

1 Paper is manufactured in rolls of total length 1 km, and on average there is one defect every 400 m. Find the probability that a roll has more than four defects.

2 On 250 successive days at a telephone exchange, the numbers of emergency calls were recorded, and were as follows:

Number of calls:	0	1	2	3	4	5 or more
Number of days:	85	92	63	8	2	0

Find the mean number of calls per day, and calculate the frequencies of the comparable Poisson distribution. Determine also the variance of the observed distribution, and compare it with that of the Poisson.

3 Numbers of particles emitted from a quantity of radioactive material were counted in 100 successive minutes, and the numbers of times that 0, 1, 2 . . . particles were emitted were as shown in the table below. Calculate the expected frequencies of the comparable Poisson distribution.

Number of particles:	0	1	2	3	4	5	6 or more
Number of occasions:	18	29	22	20	8	3	0

4 Records of deaths for each day of one year for a town gave the following numbers of days on which each number of deaths took place.

Number of deaths per day:	0	1	2	3	4	5	6 or more
Number of days (total 365):	164	135	48	12	5	1	0

State why the Poisson distribution might be expected to apply, giving any restrictions you think necessary. Calculate the mean of the distribution, and the comparable Poisson frequencies.

5 Use the Poisson distribution to give approximate answers to exercise 9 of chapter 5, which is better done by the binomial distribution.

6 A process for making electric light bulbs produces, on the average, one defective bulb in 100, when under control. Use the Poisson distribution to calculate the probability that a sample of 50 will contain (a) 0, (b) 1 or (c) 2 defective bulbs. The bulbs are delivered in batches of 1000, from which a sample of 50 is taken at random, and a batch is accepted if at most one of the 50 is defective. Calculate the probability of accepting a batch containing 50 defective bulbs. (AEB)

7 Two instruments are set to record two separate and independent sources of radiation. The numbers of particles recorded from the two sources in one second have Poisson distributions with means 2 and 4 respectively. Find the mean total number of particles per second. Find (to 3 d.p.) the probabilities that, in one second, (a) at least one particle is recorded; (b) a total of exactly three particles is recorded. (JMB)

8 A truncated Poisson distribution consists of all the terms of the standard Poisson except the first—that is, experimental results showing zero occurrences are ignored. If the probability generating function for the distribution is $Ke^{-k}(e^{kt} - 1)$, determine the value of K and the mean of the distribution. Prove that the variance of the distribution is $ke^k(e^k - k - 1)(e^k - 1)^{-2}$.

9 A swing bridge across a canal is open for two minutes while a boat passes through. Along the road in one direction pass both cars and lorries. The cars take up 10 m of road when stopped, allowing for space between adjacent vehicles, and the lorries take up 20 m. On average, two cars pass per minute, and one lorry per minute. What is the probability that the queue will extend more than 100 m from the barrier before the end of the two-minute period? (Assume that each vehicle moves independently of the others.)

Chapter 7

The normal distribution

Both the binomial and the Poisson distributions are probability distributions for a discrete variable, and for this and other reasons are of limited application only. We have noted already (page 63) how the form of the binomial tends to a limit as the number of trials increases, and in fact the Poisson also tends to a limit in the same way. When this happens, the result is a continuous probability distribution of very wide application, the normal distribution. The brief analysis which follows is considered initially in terms of a limiting binomial, but the restrictions which apply to the binomial do not all apply to the normal distribution, for reasons which are too complex to give in full.

We can picture the situation as follows. Imagine a triangular array of horizontal pegs placed so that a marble released from the top falls through them, bouncing left or right at random, as in figure 7.1. This yields a binomial distribution with $p = \frac{1}{2}$, if it is well enough constructed; other values of p could be allowed for by offsetting the pegs slightly and consistently to one side. The limiting case is now developed by extending the array downwards, and at the same time reducing the scale overall so that the scale of the distribution itself remains unaltered.

The calculations made in chapter 5 indicate that, for 25 rows, the shape of the distribution is much the same as for 16. When the number of rows increases much further still, the blocks of the histogram become so narrow that the outline may be considered to be a smooth curve. We then have a probability distribution which may be used for any large value of n with high accuracy, and whose shape is universal; only the scale needs to be adjusted in order to use it in a given problem. This is the **normal distribution** curve, and the word 'normal' is understood in this sense from now on.

The derivation of the equation of the curve is difficult, and need not concern us since we shall not make use of the equation in this book. Just for the record, it is

$$y = \frac{1}{\sqrt{(2\pi)}} e^{-z^2/2}, \qquad (7.1)$$

where z is the variate expressed in units of the SD. Since it is a probability density distribution, the equation implies that the probability that a variate will take a value in the range z to $z + \delta z$ is equal to $y\delta z$, the area under that part of the curve. As usual, the total area is unity.

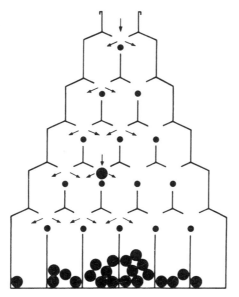

Figure 7.1   A model for generating a binomial distribution

In order to define clearly the situations in which the normal distribution may be assumed to apply, two further points may be made. First, the calculations of chapter 5 suggested the approach to a limit in the particular case $p = \frac{1}{2}$; but the derivation of the equation above makes no assumptions about p except that neither p nor q must be very small. The distribution of probabilities in the bouncing marbles experiment with offset pegs would therefore be, in the limiting case, the same—the same *shape*, that is, although the mean and SD would be different.

The second point takes the generalisation a considerable stage further. Suppose that, instead of a triangular array of pegs, we have a completely random array. (This is not easy to visualise, but it helps if you assume that both pegs and marbles are extremely small.) Then, if the number of pegs were large enough, the resulting distribution would be normal. This is a consequence of what is known as the **central limit theorem**, which states (in one of its forms) that the sum of a large number of independent variables is approximately normally distributed whatever the distribution of the individual variables. Again, the mathematics is difficult and we shall take the result without proof.

We now have a probability distribution for situations which will not yield to analysis in any other way. Consider, for example, the passage of a bullet through the air. Even if we assume that the rifle is fixed, and that the bullets are identical and have the same muzzle velocity, they will not all pass through the same hole in the target. The changes in air currents—in the limit, the collisions with individual molecules of air—will have a cumulative effect which is measurable. These collisions represent just that set of conditions required for the central limit theorem to apply, and so we should expect that the distribution of bullet holes (both horizontally and vertically) would tend to the normal. This still holds if we allow random variations in the properties of the bullets.

Biological measurements are another type in which the normal distribution can be widely applied. A man's height, for example, is determined by a large number of unrelated and unmeasurable factors, so it is reasonable to expect a normal distribution, and such proves to be the case. Some such measurements follow the normal distribution only in a modified form: if a linear dimension follows a normal distribution, then the corresponding weights (which are proportional to the cubes of the linear dimensions) will not be normally distributed. In such cases the variable has to be transformed (e.g. by taking its cube root) before the normal distribution can be used. Often the uncertainties of such situations mean that the only real justification for the use of the normal distribution is that it fits the facts reasonably well.

Dimensions of mass-produced articles in almost any manufacturing process form another wide area of application. A machine may be designed to make 3 cm nails, but the lengths will not, in practice, all be 3 cm or even all be the same—typically they will follow a distribution which is approximately normal, and whose SD is a measure of the accuracy of the machine, though there may well be some degree of skewness.

The normal distribution as an approximation to the binomial

For large values of n, the calculation of the binomial probabilities is lengthy, and if n is more than 15 or so, the normal distribution can be used as an approximation to it. If p is not near $\frac{1}{2}$, the errors are larger and the fit is better in the centre than at the tails. The first example is the 25-coin problem which has already been worked out by the binomial.

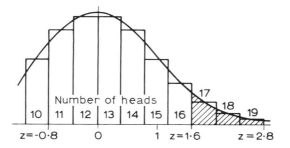

Figure 7.2   The 25-coin problem using the normal distribution

Example 1 If 25 coins are thrown together, find the probability that the number of heads is (a) in the range 17 to 19, (b) over 19, and (c) in the range 11 to 19. The data is shown in figure 7.2, which is a section of figure 5.1, not drawn to scale.

(a) We require the area under the curve (representing the probability) corresponding to the blocks for 17, 18 and 19, which is the range from $16\frac{1}{2}$ to $19\frac{1}{2}$ measured on a continuous scale. The extra $\frac{1}{2}$ on each side is the correction for continuity, which was discussed in chapter 4. We now have to express these figures as differences from the mean, measured in units of the SD.

The value of the SD is 2·5, so that $16\frac{1}{2}$ gives $(16\frac{1}{2} - 12\frac{1}{2})/2\frac{1}{2}$, which is 1·6, and $19\frac{1}{2}$ likewise gives 2·8. The variates expressed in this way are called **standardised variates**: they are shown in the right-hand part of figure 7.2, and this information is all that is required to find the probability.

The concept of a standardised variate is an important one. Essentially we use it to express a wide variety of units and measurements in a single form: the number of standard deviations between a particular variate and the mean of its distribution. It is not only an arithmetical alteration— it should affect our thinking processes as well. Just as the linguist learns to think as well as to speak and write in a foreign language, the statistician must learn to think in units of SD's whenever he uses the normal distribution, and that is often.

Mathematically, the standardised normal variate is

$$z = \frac{x - \mu}{\sigma}, \tag{7.2}$$

where x is the variate, and μ and δ are the mean and SD of its probability distribution. With experience, you will be able to use an observed value of z to form a rough conclusion before consulting the tables for a precise one—for example, if the value of z was near 3 one could say at once that this was a rare case, one for which the probability was very small indeed.

Because the normal distribution is a single curve, the areas under it can be tabulated for any application, and this is done in Table 5 on page 179. The figures in the table give the areas under the curve from minus infinity (i.e. the left-hand end) to the positive value of z, the standardised variate. They start, therefore, with 0·5000 at $z = 0$, and end with figures near 1 as z becomes very large. The area shown shaded in figure 7.2 is the probability that a random variate should be in the range 1·6 to 2·8, which is the difference between the areas to the left of $z = 1·6$ (the probability of less than 17 heads in 25 throws), and to the left of $z = 2·8$ (less than 20 heads). From the table, these areas are 0·9452 and 0·9974 respectively, so that the difference is 0·0522. The exact value from the binomial is 0·0518, so that the error is less than 1 %.

(b) The probability of getting over 19 heads is the area under the tail beyond $z = 2·8$, which is the amount by which 0·99744 falls short of 1, 0·00256. The actual figure is 0·00204, not a large difference but, in proportion, much greater than before—just over 25 %.

(c) The area from $-0·8$ (the standardised form of $10\frac{1}{2}$, the lower side of the 11 block) to 2·8 is worked out rather differently because it straddles the mean. The easiest way is to work out the area from $-0·8$ to 0 (the same as 0 to 0·8, because of the symmetry of the curve) which is 0·2881, and to add to this the area from 0 to 2·8, which is 0·4974, giving 0·7855. An alternative method is to find the area to the left of $-0·8$ by subtracting the tabulated figure for 0·8 from 1, and subtracting the result from the tabulated figure for 2·8 : $0·9974 - (1 - 0·7881) = 0·7855$.

Example 2 A pack of cards is cut 10 times; find the probability of cutting exactly 4 spades. This also has been worked out by the binomial, and two factors indicate that the normal approximation can only be used to give a very rough answer: 10 is small for the number of trials, and p is not near $\frac{1}{2}$. The SD is $\sqrt{(10 \times \frac{1}{4} \times \frac{3}{4})} = 1·37$, and the mean is $2\frac{1}{2}$, so the standardised forms of $3\frac{1}{2}$ and $4\frac{1}{2}$ are 0·73 and 1·46. This gives a probability of $0·9279 - 0·7673$ or about 0·16; the actual figure is 0·15 so the accuracy is quite good. But the figure for 1 spade (the same standardised variates as before, only negative) is also 0·16, which is rather further from the actual value of 0·19. The figure for 6 spades is 0·0125, which compares with 0·0166; and one can also give a 'probability' of 0·0125 of cutting -1 spade!

The normal distribution as an approximation to the Poisson

For large values of the mean, the calculation of the Poisson probabilities becomes tedious after the first few terms of the series. The normal distribution gives a fair approximation to it for values of μ greater than about 15, the errors being greatest in the first few terms. The errors in the tail are also large in proportion to the size of the probabilities. To give a rough indication of what happens, some of the figures for $\mu = 9$ are given below.

No. of observations	Poisson prob'y	Normal approx'n
0	0·000 1	0·001 5
3	0·015 0	0·018 3
6	0·091 1	0·080 7
9	0·131 8	0·132 4
12	0·072 8	0·080 7
15	0·019 3	0·018 3
18	0·002 9	0·001 5
21	0·000 26	0·000 04

Examples of the use of the normal distribution

Example 3 A machine produces ball bearings of nominal diameter 0·500 cm with tolerances of $\pm 0·0020$ cm (i.e. the actual diameters have to be in the range 0·4980 to 0·5020 cm). In a long run, 5% are rejected as oversize and 1% as undersize. Assuming that the output is normally distributed, find the mean and SD. If the machine is reset so that the mean of the output is 0·500 cm, and if the SD is the same as before, what will be the total percentage of rejects?

Table 3 gives the percentage points of the normal distribution directly: these are $z = 1·645$ for a 5% tail, and 2·326 for a 1% tail. The sum of these is equivalent to the difference between the limits, which is 0·004 cm. Hence the SD is 0·004/3·971 which is 0·001007 cm. The mean is 0·4980 + 2·326 × 0·001007 = 0·50034 cm.

If the mean is reset to 0·500 cm, each of the limits is now 0·002/0·001007 = 1·99 SD's from the mean. The proportion of rejects is therefore 0·0233 on each side, giving a total of 4·66%—a considerable improvement on 6%.

Example 4 Ten thousand men's raincoats are to be made to a particular design in different sizes to suit men in each 5 cm range of height. The adult male population is known to have mean height 169 cm and SD 8 cm. How many of each size should be made? A tabular calculation is convenient.

Variate	Standardised variate	Cumulative probability	Probability for each size range	Number in each size range
140	−3·625	0·000 1	0·001 2	12
145	−3·000	0·001 3	0·007 5	75
150	−2·375	0·008 8	0·031 3	313
155	−1·750	0·040 1	0·090 2	902
160	−1·125	0·130 3	0·178 2	1 782
165	−0·500	0·308 5	0·241 3	2 413
170	0·125	0·549 8	0·223 6	2 236
175	0·750	0·773 4	0·142 0	1 420
180	1·375	0·915 4	0·061 8	618
185	2·000	0·977 2	0·018 5	185
190	2·625	0·995 7	0·003 7	37
195	3·250	0·999 4	0·000 5	5
200	3·875	0·999 9		9 998

The total is not exactly 10 000 because of the small but measurable probability that a man is outside the 140–200 cm range of heights. The two odd ones would have to have garments specially tailored, and in practice others would doubtless have to do so as well, since batches of 5 and 12 would not be justifiable on grounds of cost. The remaining figures would probably be rounded off—the actual frequency distribution of the first 10 000 enquirers would not be precisely the expected distribution, and might even be very different if marketing conditions were not as forecast.

Example 5 A packing machine fills paper bags with powder which costs 10p per kg. The bags have to contain at least 1 kg, and the output from the machine is normally distributed with SD 0·01 kg. All the output is checked, and underweight bags are made up by hand at a cost of 3p each. To what value should the mean weight of the bags be set, and what is the corresponding average cost of a bag?

This cannot be done by an equation, and so trial and error is necessary. (It is not, therefore, a type of problem met in examinations, but is typical of an actual manufacturing problem.) If the mean is set at 1 kg, half the output will be underweight and so the costs of making up will be high. But if it is set at 1·05 kg, so that underweight packs are the insignificant few which are more than 5 SD's below the mean, a lot of powder will be wasted in overweight bags. The problem is to find the optimum, and again a tabular layout is best.

Mean weight g	Cost of powder p	Fraction underweight	Cost of making up	Total mean cost, p
1·01	10·1	0·158 7	0·476	10·576
1·02	10·2	0·022 8	0·068	10·268
1·03	10·3	0·001 3	0·004	10·304
1·04	10·4	0·000 0	0·000	10·400
1·025	10·25	0·006 2	0·019	10·269
1·022 5	10·225	0·012 2	0·037	10·262

The first four lines gives the general picture, and indicate that somewhere between 1·02 and 1·03 will give a minimum cost. The figure for 1·025 is calculated next, and the cost is almost the same as for 1·02, which means that the optimum must lie between them. So the mean is set to 1·0225 kg, and the resulting mean cost is 10·262p per bag. This might be reduced still further, but only by one or two thousandths of a penny. With this setting, about one pack in 80 will have to be made up to full weight.

The properties of the normal distribution curve

Because of the wide use of this distribution, a knowledge of its main features is useful.

(a) The curve is symmetrical, with its mode and median both equal to the mean.

(b) The points of inflexion (that is, the points where the gradient is numerically a maximum) are at the points where $z = \pm 1$. (This can be proved without difficulty from the equation of the curve.) Using this fact, a quick sketch can be drawn to fit any normally distributed data.

(c) From Table 5, 68·3% of the distribution lies between the ± 1 SD points. This provides the justification for the simple test already used, that approximately $\frac{2}{3}$ of the distribution lies between these points, or alternatively that the semi-intersextile range provides a rough estimate of the SD.

(d) Likewise, 95·4% of the distribution lies between ± 2 SD's of the mean. The approximate figure of 95% is often used here, and should be memorised.

(e) Finally, 99·73% of the distribution lies between ± 3 SD's of the mean; or, in other words, only about one four-hundredth lies outside this range. The important part of the curve therefore lies between these points for most purposes.

Summary for quick reference

The probability density curve for the normal distribution may be used as an approximation to the binomial if $n \geqslant 15$ for values of p around $\frac{1}{2}$, and for rather higher values of n if p is near (but not too near) 0 or 1. Errors are high towards the tails. It may be used as an approximation to the Poisson if $\mu \geqslant 15$, but again, not towards the tails. In order to use the normal distribution, the mean and SD must be known, and the variates are then expressed in standardised form using the equation

$$z = \frac{x - \mu}{\sigma}$$

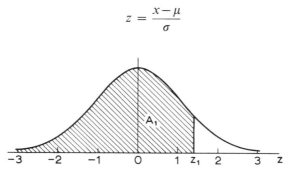

Figure 7.3   The normal probability distribution curve

The probability that a random variate will be less than a specified value of z is given in Table 5. This is illustrated in figure 7.3, the area A_1 representing the probability. The probability that a random variate will lie between z_1 and z_2 is equal to $A_2 - A_1$. Table 5 gives the probabilities for positive values of z; for negative values, the figure for the corresponding positive value is subtracted from 1.

Exercises

1 A certain make of television tube is found to have a mean life of 2400 hours, normally distributed with SD 300 hours. What is the probability that a random tube will (a) fail under 1500 hours; (b) last more than 3000 hours; (c) fail between 2200 and 2600 hours.

2 Thirty-six fair coins are thrown. Use the normal approximation to the binomial (allowing for the continuity correction) to find the probabilities of (a) more than 22 heads; (b) more than 25 heads; (c) a number of heads in the range 16 to 20 inclusive.

3 Resistors of nominal value 50 kilohms are found to be on average 49·8 kilohms, normally distributed with SD 0·6 kilohms. If the specification requires the resistance to be in the range 50 ± 1 kilohms, what proportion are unacceptable? If the manufacturing process is altered so that the mean becomes 50 kilohms, what proportion will then be unacceptable?

4 A piece of radioactive material emits particles with a mean interval of 0·0625 seconds between successive emissions. Use the normal approximation to the Poisson distribution to find the probability that, in one second, (a) more than 20, (b) more than 25 particles will be emitted.

5 Vehicles pass independently along a road at an average rate of five per minute. Use the normal approximation to the Poisson distribution to estimate the probability that, in a given 10-minute period, more than 60 vehicles will pass.

6 A so-called 'objective test' consists of 100 questions with four alternative answers to each, of which only one is correct. Find the probability of scoring 40 or more correct answers by random guessing.

7 An 'intelligence quotient' is normally distributed over the whole adult population, with mean 100 and SD 15. What percentage of the population have IQ's (a) over 150, (b) over 140, and (c) between 90 and 110? (The figure is taken to the nearest unit.)

8 The maximum flow at the time of the annual floods of a certain African river is found from records taken over 80 years to be approximately normally distributed, with mean 6300 m^3/s and SD 1900 m^3/s. What is the flow which will be exceeded on average once every ten years? (This is known as a 'ten-year flood'.) What are the corresponding figures for a hundred-year flood and for a thousand-year flood?

9 A high-jumper finds that he can clear a height of 1·5 m three times in every five jumps, and 1·6 m once in every five jumps. Assuming that the height he actually clears is normally distributed, find the height (as a multiple of 1 cm) which he can expect to clear once in every 100 jumps.

10 Components produced by an automatic lathe are checked against two gauges of 1·995 cm and 2·005 cm: they are rejected if they pass through the first or fail to pass through the second. The observed

failure rates are 3.6% undersize and 1.2% oversize. Find the mean and SD at which the machine was operating, assuming a normal distribution, and also the total percentage rejected if the mean were reset to 2.000 cm, the SD remaining the same.

11 The probability of six or more heads in 10 coins is about 0.38, using the binomial distribution. Compare the probabilities of getting 60 or more in 100, and of 300 or more in 500, using the normal distribution.

12 In packing bags of powder which are nominally 1 kg in weight, it is required that the proportion underweight shall be less than 2%. The packing machine produces a normally distributed output with SD 7.5 g; to what value should its mean be set?

13 A machine packs bags of sugar of nominal weight 1 kg, and the bulk sugar costs 10p per kg. The packing machine can operate on two settings: in each case it produces a normally distributed output. If it works at the slow speed, the SD of the output is 0.005 kg, and the operating cost per bag is 0.2p; at the fast speed, the SD is 0.01 kg and the cost 0.1p. If 1% are permitted underweight, find which is the more economical setting.

14 A powder costing 15p per kg is packed into 1 kg bags. The packing machine produces a normally distributed output with SD 0.005 kg. No underweight bags are permitted, and those which come from the machine underweight have to be made up to weight by hand, at a cost of 2p each. Determine by trial the optimum setting for the mean.

Chapter 8

Samples and sampling distributions

The process of reaching conclusions by taking samples is one which is basic to each one of us—we started it in the first year of our lives when a sample of one mouthful of a particular type of baby food made us decide that a second mouthful was either to be taken eagerly, or to be rejected. When considered in terms of probability, the method is known as **statistical inference**: it is the art of drawing conclusions about a population as a whole by examining or testing a part of it only.

Sampling may be made necessary for one of three reasons:

1. The testing process may be destructive, so that, if all the population were tested, nothing would be left. Wine-tasting is one example.

2. The population may be too large to test economically; examples are many types of manufacturing processes. There may be advantages in sampling even if all the population could be tested, because smaller numbers means better control over the testing procedure.

3. The population may be indefinable or infinite: for example, the set of all possible measurements of the velocity of light, under given experimental conditions.

In this chapter we shall first take a look at the ways of choosing a sample, and then consider a typical calculation based on sample measurements, the estimation of a population mean.

Measurements made on a sample (mean, median, SD, etc.) are known as **statistics** (singular **statistic**—this was the technical use of the word referred to in chapter 1); the corresponding population measurements are called **parameters**.

Taking samples

The aim in sampling is to obtain a fully representative sample. This is obvious enough, but is often difficult except in experiments of an academic nature. Some bias is usually inevitable, and the best one can do is to keep control of its extent. Only an introduction to this complex subject can be given here, and it must start with a word of caution: pitfalls for the unwary abound in all branches of statistics, but nowhere are they more frequent than in sampling procedures.

Differences in calculated figures, whether as a result of bias or because of the inherent loss of information involved in taking a sample, are known as **sampling errors**. The word 'error' is used in the sense of a difference between an observed figure and the value that figure would take if full data were available.

In almost all sampling procedures there is some way of ensuring that no discretion is allowed to the human agent involved: for example, one typical method is to take every tenth item. We must make sure that this itself does not introduce bias, as could happen if there was some cyclic pattern in the population. Further, the first item chosen must not always be the first one of a batch—in a survey involving interviews at every tenth house, the interviewer must not always start at the end house of a road, or there would be a bias towards corner houses.

A selection process in which every item in the population has an equal chance of being chosen generates a **simple sample**. This is, in fact, somewhat rare because, once an item has been chosen, it is usually ineligible for selection again. A sample from a pack of cards is a simple sample if one card is drawn at a time and then replaced (and the pack reshuffled) before the second card is drawn, and so on until the sample is complete. The phrase 'random sample' usually implies, in practice, a sample drawn in an unbiased way without such replacement.

In most of the calculations later on, the various formulae are derived on the basis of simple samples, and then applied to situations where the samples are drawn without replacement. This is only legitimate if the population is very much larger than the sample, and usually this is the case. Where it is not, a suitable correction factor can be applied, as in equation 8.3.

When a population is small enough to be listed, a sampling method based on the use of random digits is best. Tables of such digits (usually generated nowadays by computer) have been constructed, and one is included as Table 12 at the end of the book. Suppose, for example, a sample of 50 is required from a listed population of 4357 items. Each item is first given a serial number. The digits are then grouped by fours in sequence, and any combination of digits outside the range 0001 to 4357 is ignored. This continues until 50 suitable four-digit numbers have been found, and the items with these serial numbers form the sample. This, in fact, generates a simple sample, but in practice a duplicated number would probably be passed over anyhow.

For larger or less definable populations, a systematic sampling method is the best approach to randomness. A random number between 1 and n is chosen to determine the starting point, and thereafter each n'th item is chosen. As mentioned before, there must be no cyclic pattern (at least if the cycle length is any multiple or factor of n) in the population. But it must be possible to order the population in some non-discretionary way—this is easy enough for houses in a street, but may be more difficult for farms scattered over a wide area.

When a population can be unambiguously classified into categories, the method can be improved slightly. For example, if a sample of 50 is required from a firm's payroll, and if 62% of the employees are men, then the right proportion of men to women will be obtained by choosing 31 men and 19 women. This is known as **quota sampling.**

Economic considerations may require modifications which tend to increase sampling errors. For example, if a 1% sample from a large town is required, an area of the town thought to have a representative cross-section of the different types of housing, and large enough to contain 10% of the town's population, might be used. From this area a 10% sample would be taken. This is clearly more economical in manpower, because the interviewers do not have to walk as far. But it is no longer a random sampling method for the whole town.

As a general rule, money is usually better spent on a small sample with the very minimum of bias than on a larger sample of doubtful randomness. But very small samples must clearly be avoided.

Finally, some of the pitfalls.

(a) An element of subjective choice. The mass-media technique of interviewing people at street corners is perhaps the commonest example. In this case the errors (even if there is no deliberate slanting) are obvious. But they are still considerable, if less obvious, when there is a choice apparently without bias. Anyone doubtful of this should try writing down a 'random' list of 100 H's and T's to represent the results of a coin-spinning experiment. Then look at each of the 96 groups of five consecutive letters, and count how many of them are all H's or all T's—in a genuinely random list for which $p = \frac{1}{2}$ for each, the expectation of the number of such groups is six. Experience shows that, in most lists written down 'at random' (at least by people who do not know that about six sequences of five are to be expected in a random list), the actual number is less than six. Also, those who try the hardest to be random are generally the least random of all!

(b) Incompleteness of a random selection procedure. In most opinion or factual surveys, this is where perfection has to be left far behind. An interview often cannot be carried out, either because no one is at home after repeated calls, or because the person refuses to answer. In the decennial Census of the United Kingdom, this second problem is to a large extent overcome by the force of law, but in most cases it is a major source of bias. Postal surveys, though far more economical than interviews, are particularly prone to this type of error.

(c) Substitution. If, in a 10% house-to-house survey, someone is out, then surely the house next door is just as random? Certainly not, because the proportion of households with working mothers will be underestimated.

(d) Selection from a non-representative group. This is a fertile source of bias, and one example has already been given in area sampling. The classic example is the 1936 survey which incorrectly predicted that Landon would defeat Roosevelt in the American Presidential election. The survey was a postal one, which doubtless accounted for part of the error, but the main reason was that telephone directories were used as the mailing list. Poorer people were therefore under-represented, and it was from them that Roosevelt got most of his support.

(e) Incompleteness of data about the population itself. The organisers of the decennial census, for example, take endless pains over this, but their figures for vagrants cannot possibly be accurate.

(f) Errors in the testing, measurement or interview procedure. Even in the simplest matter, it is difficult to be fully objective. Is a dimension just over the tolerance allowed, or just within? Individuals may give different answers. In any form of interview, sources of error abound: questions can be loaded, imprecise, ambiguous or leading, and answers can be guessed, exaggerated, rounded off, unconsciously mis-leading or simply untrue. For this reason alone, any survey reports about matters which are essentially personal should be treated with considerable suspicion. Those who 'don't know' or are undecided may add more bias, which is particularly important in political sur-veys, where a 1% error may be sufficient to tip the balance of the predicted result of an election.

In all that follows, it is assumed that a sample is chosen randomly, and where in practice this is known not to be entirely true, the results of the investigation must be treated with a corresponding amount of caution.

Sample mean and population mean

The first parameter to estimate for any population is usually its mean, and the sample mean forms an obvious point estimate. But a point estimate is not enough: we must determine, if possible, three further things about it. First, we must ensure that the estimate has no tendency to err consistently on one side or the other—that is, it should be an **unbiased estimate**. Second, we must find out how close to the true value the estimate is, and this will be in the form of the SD of its probability distribution. Finally, we must know the form of this distribution.

A sample of given size from a given population will, of course, vary in its content. A statistic of it will therefore also vary and, since its distribution follows the laws of probability, we can find what that distribution is. The probability distribution of a sample statistic is called a **sampling distribution**. (Note carefully that it *is* a probability distribution, even though its name may seem to imply that it is an observed one.)

The first condition mentioned above, referring to the estimation of a population mean, can now be expressed as follows: the mean of the sampling distribution of the sample mean must be the population mean, if the sample mean is to be an unbiased estimator. (We can express this requirement also by saying that the expected value of the sample mean must be the population mean.) This is in fact so, though it should not be taken as a precedent when considering other sample statistics. The proof which follows is not particularly easy, and may be omitted at a first reading.

Imagine a row of n boxes, each to be filled by a random member of the population to form the sample. The item which fills the first box has a probability distribution which is identical to the frequency density distribution of the population. (If you find this difficult, think of a simple numerical case, such as a population of one hundred 12's, three hundred 13's and two hundred 14's: the probability distribution for a sample of one is 1/6 for 12, 3/6 for 13 and 2/6 for 14, the same as the population frequency density distribution.) The mean of the probability distribution is therefore the population mean. The same is true for the item which fills every other box, assuming a simple sample; if there is no replacement, the distribution will alter slightly but not in a biased manner. The mean of the probability distribution of the sample mean is the sum of the means for each member of the sample divided by the number in the sample, and since the mean for each member is the population mean, the overall mean will be the population mean also. The sample mean is therefore an unbiased estimator.

The distribution of a sample mean

Next, we have to find the SD of the distribution of the mean. The SD of a sampling distribution is known as a **standard error** (SE), mainly for historical reasons—though it also helps to avoid confusing the SD of the sample itself with the SE's of its statistics. The SE of a sample mean will be denoted by σ_n. Consider a population having mean μ and SD σ, from which samples of size n are drawn. Then σ_n is given by the formula $\sigma_n = \sigma/\sqrt{n}$. So, the larger the observed sample mean, the better the estimate of the population mean, which is what common sense predicts. The proof which follows may again be omitted at a first reading.

The mean of a sample is the sum of the variates divided by the number of them, and so, in order to find the variance (or SE) of the mean, we must first calculate the variance of the sum of two or more independent random variates.

In a simple form, the problem is this. Suppose a certain component in a machine consists of two blocks of steel fixed together, the first being nominally 6 cm thick with SD 0·003 cm, and the other 8 cm thick with SD 0·004 cm. Then, if two batches of 100 are paired off randomly, giving pairs with a mean total thickness of 14 cm, what will be the SD of the total thickness?

The mathematical way of stating the same thing is as follows. We have two distributions (not necessarily normal) of n random variates, having SD's s_x and s_y, and typical deviations from the means d_x and d_y. They are added together randomly term by term, and we must find the SD s of the resulting distribution. Then

$$ns_x^2 = \Sigma d_x^2$$

$$ns_y^2 = \Sigma d_y^2$$

$$ns^2 = \Sigma(d_x + d_y)^2 = \Sigma d_x^2 + \Sigma d_y^2 + 2\Sigma d_x d_y$$

The difficulty is to evaluate the last term, since we do not know which d_x goes with which d_y. It is in fact zero, and the simplest way of seeing this is as follows: for a given d_x, the corresponding d_y is as likely to be positive as negative since it has a mean of zero, and so, taken overall, the positive terms will cancel out the negative ones.

Rather more rigorously, we can allow for the randomness of the pairing procedure by taking all possible pairs. (Since this is n times as many pairs as we want, we would have to divide by n at the end if the sum were not zero.) A typical d_x is thus paired off with every d_y in turn, and for one such d_x the sum of the products $d_x d_y$ is $d_x \Sigma d_y$. But this is zero since Σd_y is zero, and so the whole term $2\Sigma d_x d_y$ is also zero. It follows that

$$s^2 = s_x^2 + s_y^2 \qquad (8.1)$$

In other words, the variance of the sum of two random variates is the sum of the variances of the original distributions.

Strictly speaking, what we have found is not an exact expression for the variance of the sum—this cannot be determined because it depends on the actual pairings, which are random. We have, instead, an unbiased estimator. But, although the distinction is important in many cases, it may be neglected here.

For three random variates, $s^2 = s_x^2 + s_y^2 + s_z^2$; this can easily be seen by adding two of them and then adding the third to the result. The proof can similarly be extended to any number of variates.

We can also note in passing that the variance of the *difference* of two random variates is also equal to the sum (and not the difference) of the variances. The proof follows the pattern above, the only alteration being that $\Sigma(d_x - d_y)^2$ when multiplied out gives a negative sign for the product which is zero anyhow. This result will be required in chapter 10.

In the numerical example quoted earlier on, the SD of the total thickness is 0·005 cm—more than either of the original SD's but less than the sum of them, as might be expected.

A sample of n items is, as we noted earlier, n independent random variates, which are identically distributed because they come from the same parent population. Again, it is assumed that the population is sufficiently large compared with the sample for the lack of replacement not to affect the distribution significantly. The variance of the sum of these n items is therefore $\sigma^2 + \sigma^2 + \sigma^2 + \ldots + \sigma^2 = n\sigma^2$, and its SD $\sqrt{(n\sigma^2)} = \sigma\sqrt{n}$. Since the mean is the sum divided by n, the SD is reduced by the same factor to give the **SE of the mean**, which is therefore given by

$$\sigma_n = \sigma/\sqrt{n} \qquad (8.2)$$

If it cannot be assumed that the population is much larger than the sample—say by a factor of at least 20—the following alternative form may be used. For the proof, a more advanced text must be consulted.

$$\sigma_n^2 = \frac{\sigma^2}{n} \times \frac{N-n}{N-1} \qquad (8.3)$$

Finally, we consider the form of the distribution. The parent population may, of course, have any distribution, but it is a remarkable fact—too difficult, alas, to prove here—that, except for very small values of n

drawn from a population which is far from normal, the distribution of the sample mean is very close to the normal. (In exercise 10 at the end of the chapter, a method of verifying the result experimentally is suggested.) This fact is a consequence of the central limit theorem, which was mentioned in chapter 7. There is, however, one further condition—the SD of the parent population must be established independently of the sample. If the sample itself is the only data for estimating it, a modified form of the normal distribution is used: this is the *t*-distribution, discussed in chapter 9.

Everything is now known about the sampling distribution, and the next stage is to decide how to express the conclusions. Apart from the point estimate, this can be done in two ways: by giving a confidence interval, or by using a significance test. These are the subject of the next chapter. The example used initially will not be the estimation of a population mean, because of the difficulty over the SD mentioned above. Instead, we shall consider a population consisting of items of two types, such as supporters of two political parties, and the problem of estimating the proportion of each from sample data.

Exercises

Sampling and survey methods

1 Discuss the possible methods and difficulties of taking a random sample in the following situations.
   (a) A 5 % sample of the children between ages 11 and 16 in a particular town.
   (b) 10 mice out of a group of 100, for experimental purposes.
   (c) A 10 % sample of the area of a large field, taken to estimate the numbers of different varieties of wild flowers.
   (d) 100 ml of wine from a cask.
   (e) 10 litres of water from a river, for pollution tests.
   (f) 100 g of powder, from a drum with an opening in the top.
   (g) 1 kg of topsoil from a field.
   (h) 1000 radio listeners or television viewers.
   (i) 1000 users of a given make and model of car.
   (j) 1 kg of coal from a delivery lorry.
   (k) 100 unemployed men from a medium-sized town.
   (l) 100 kg of surface dust and rock from the moon.
2 A manufacturer of a new soap powder wishes to predict the likely volume of sales in a town. Four schemes, as below, are proposed for

selecting people for a questionnaire. Discuss the merits of each and choose one, explaining why you think it the best.

(a) Take every 20th name on the electoral register of the town.

(b) Choose people entering a supermarket, ensuring that the numbers in each sex, age and social class category are proportional to the number in the population.

(c) Select houses at random from a town plan and interview one person from each.

(d) Choose at random one name from each page of the telephone directory and ring them up. (OL)

3 Discuss the possible sources of bias in answers to the following survey questions, and where appropriate criticise and re-word the questions themselves.

(a) What is your age?

(b) What is your income?

(c) Are you married? (Consider men and women separately.)

(d) How often do you have a bath?

(e) State which of the following newspapers or magazines you read regularly. (Make up your own list, covering serious, popular and worse, and comment on each.)

(f) If there were a General Election tomorrow, for which party would you vote?

(g) At the last General Election, for which party did you vote? (Note that this is an example of a very useful kind of survey question, one whose answers can be compared with available and accurate information as a check on possible bias.)

(h) (To married men.) Do you reveal your income to your wife? (This also is open to checking, if a corresponding question is asked of married women; the check will be direct if the questions are asked of married couples, or statistical if asked of separate groups.)

(i) Are you in favour of Civil Rights?

(j) Are you in favour of preserving the beauties of the countryside?

(k) Do you approve of the addition of fluoride to the public water supply in order to reduce tooth decay, particularly among children?

(l) Do you agree with compulsory medication by the addition of fluoride to the public water supply?

Variances of combined distributions

4 A colony of a particular variety of sea-bird is known to consist of approximately equal numbers of males (mean mass 180 g with SD

20 g) and females (mean mass 210 g with SD 30 g), the masses in each case being normally distributed. Find (a) the probability that a bird selected at random will have a mass of less than 150 g; (b) the probability that a mating pair (assumed random) will, together, have a mass of less than 320 g.

5 The amount of luggage that aircraft passengers take with them is distributed with mean 20 kg and SD 5 kg. A certain type of aircraft carries 100 passengers. What is the probability that the total amount of the passengers' luggage exceeds 2150 kg? To what extent does your answer depend on the distribution of individual masses being normal, and what other assumptions have you made? (OL)

6 The distribution of breaking loads for strands of rope is approximately normal, with mean 20 units and SD 2 units. A rope is assumed to be made up of 64 independent strands, and to have a breaking load which is the sum of the breaking loads of all the strands in it. Find the probability that such a rope will support a weight of 1300 units. The manufacturers wish to quote a breaking load for such ropes that will be satisfied for 99 % of ropes: determine what breaking load should be quoted. (JMB)

7 A doctor working in a clinic finds that the consulting times of his patients are independently normally distributed with mean 5 minutes and SD 1·5 minutes. He sees his patients consecutively with no gaps between them, starting at 10 a.m. At what time should the tenth patient arrange to meet a taxi so as to be 99 % certain that he will not keep it waiting? If the doctor sees 22 patients in all, what is the probability that he will finish before noon? (OL)

8 A rocket consists of three stages which are fired in succession. For the first, the velocity imparted is a normal variate with mean 2600 m/s and SD 80 m/s; the second and third stages give increases of velocity which are also normal variates, each with mean 2850 m/s and SD 60 m/s. For the third stage to go into orbit, a velocity of 8000 m/s is required; what is the probability of failing to achieve orbit?

9 A cylindrical plug has to fit into a socket, with the clearance (i.e. the difference between the radii) in the range 0·001 cm to 0·009 cm. The radius of the plug is a normal variate with mean 2·452 cm and SD 0·0008 cm, and the radius of the socket is a normal variate with mean 2·457 cm and SD 0·0015 cm. If the components are paired randomly, what proportion will be rejected? If the mean of the socket radius is wrongly set to 2·456 cm, what proportion will then be rejected?

Standard Error of a sample mean

10 Use the table of random digits (Table 12) to derive an experimental approximation to the probability distribution of a sample mean or a sample sum. The table gives directly 200 sets of 10 digits, and rather more if used with a little ingenuity; if further digits are required, which will be so if more than 10 or so people share in the calculation, an old telephone directory may be used. (But used with discrimination —the penultimate digits of the numbers are best.) The probability distribution for the population has a mean of $4\frac{1}{2}$, and is, of course, rectangular in form; the variance will be found to be $8\frac{1}{4}$. The sample sum for a sample of 10 will therefore have a mean of 45 and variance $82\frac{1}{2}$, and by the central limit theorem will be approximately normally distributed; the sample mean will have the same distribution reduced by a factor of 10. About 1000 sets of 10 random digits will be found to produce a distribution very close to this.

Chapter 9

The nature of statistical decisions

In the first chapter of this book, a topical example of the need for decision-making to be based on sound statistical analysis was mentioned—the question of whether educational standards are affected by the change from a selective to a comprehensive system. We now have a sufficient basis of statistical knowledge to be able to consider problems of this kind, though at first the examples will be rather simpler. Since the procedure is based entirely on the laws of probability, the decisions we reach are in the form of probabilities also. We can never say with absolute accuracy, 'Statement X is true'—only that it is probably true, or almost certainly true. In this chapter we shall develop more precise forms of conclusion than these somewhat vague ones, but they will still be in terms of probabilities. The distinction between statistical decisions and absolute decisions could hardly be stated better than by Bishop Joseph Butler, over 200 years ago: 'Probable evidence is essentially distinguished from demonstrative in this, that it admits of degrees.'

At first sight a statistical conclusion might seem to be much less useful than an absolute proof. And so it is, in the tidy world of the mathematician where everything is known with absolute accuracy. But in the real world, where almost nothing is known with such accuracy and random variables are the rule rather than the exception, statistical decisions are the norm. It is worth quoting a mathematician, James Clerk Maxwell, on this: 'The only true logic for this world is the calculus of probability, the only mathematics for practical men.'

One rather special example is sufficient to establish the point beyond question. The Second Law of Thermodynamics is a physical law which is perhaps more nearly indisputable than any other in the whole realm of science, and it is a statistical law. It is, for practical purposes, certain, just because the probability of its being violated is utterly remote: as remote, for example, as the probability of a random typist (such as a monkey) typing out the works of Shakespeare without a mistake.

As an illustration of the ways in which a statistical conclusion can be expressed, consider the problem of a large bag containing marbles which are identical except that each one is either red or black. From these, a random sample is drawn. From the proportion of reds in the sample (say 50 out of 100), we can draw some conclusions about the proportion of reds in the bag. Although the example concerns a proportion the same method will be used for a mean and, indeed, for any of the constants connected with a population. These constants are known as **parameters**, as mentioned in chapter 8. The conclusions may be expressed in three distinct ways.

1 We can make a **point estimate** of the proportion: on this data the best estimate is 0·5. This is the most common form of estimate at the layman's level, but the least useful.

2 We can give a probable range of values in which the proportion lies. If there are 10 000 marbles in the bag, then 50 reds in 100 implies that the number of reds is at least 50 and at most 9950, so that the extreme limits of the proportion are 0·005 and 0·995. But a figure close to these actual limits is most unlikely, and we can give more realistic limits in the form of probabilities, e.g. 'The proportion of reds lies between 0·37 and 0·63 with 99 % confidence.' (The figures given here must be taken on trust for the moment.) The range of values is known as a **confidence interval**, and the ends of the range as **confidence limits**. Typical confidence levels are 95 %, 99 % and 99·9 %, depending on the requirements of the problem; the higher the confidence, the wider is the interval. The precise meaning of the phrase '99 % confidence' will be considered later; for the moment we can think of it in the rather imprecise way we sometimes use '99 % certain' in everyday speech.

3 If we have grounds for believing that the proportion has a particular value, we can test whether the proportion in the sample is sufficiently close to it. The suggested value, a theory put forward for investigation, is called a **null hypothesis**. With an experimental result of 50 reds in 100, any postulated proportion in the range 0·37 to 0·63 (say) is sufficiently close to the observed value for the difference not to be considered 'significant'—not to be large enough to be unusual. A proportion outside those limits would, correspondingly, give a 'significant' difference between the postulated and observed values. A difference as great as this would, in fact, occur in only 1 % of trials of the experiment—that is, if the actual proportion were 0·63, the probability of a result as far from this as 0·5 is only 1 %. The observed result is then

said to be 'significant at the 1 % level'. (We can now see the link between the two tests, since the range above is the confidence interval for the percentage complementary to 1 %.)

A test of this kind is called a **significance test**, and the levels commonly used are those corresponding to the standard confidence levels, i.e. 5 %, 1 % and 0·1 %. The range in which the observed proportion can lie for the test to show a significant result is called the **critical region**; it is the complement of the confidence interval when the associated percentages are also complementary. A confidence interval can thus be defined as the range of values for which the corresponding significance test gives a non-significant result, and, from a logical viewpoint, this is probably the best way of looking at the two tests. But the idea of a significance test is probably the more difficult of the two, and so it is taken here in second place. Once the significance test is understood, you can go back to the confidence interval in order to work out the implication of the percentage associated with it, which was left undefined earlier on.

Note that the significance level is not the probability that the observed result would occur on the basis of the null hypothesis; it is the probability that a result *at least as far from the predicted value as* the observed result would occur. We can take two simple cases to make things clearer.

Consider first a null hypothesis that the proportion is 0·6. This is in the 99 % confidence interval and so, on this basis, a sample with 50 reds in 100 would not be unusual. A proportion of 0·6 is therefore 'not significant at the 1 % level'.

Next, consider a null hypothesis that the proportion is 0·75, which is outside the 99 % confidence interval, and thus within the 1 % critical region. The observed proportion is now further removed from the postulated proportion, and the probability that so great a difference would occur is less than 1 %. The observed proportion is thus 'significant at the 1 % level', and the evidence against the null hypothesis is fairly strong.

An alternative definition of the confidence interval may conveniently be mentioned at this point. The confidence level is not (though it is sometimes wrongly taken to be) the probability that the unknown parameter lies in the stated interval—the parameter is fixed even though unknown, and so a probability (either in terms of a frequency or a symmetry definition) cannot be given for it. We can, however, take the interval itself as a random variable, and determine the probability that it contains the fixed but unknown parameter. This probability is the confidence level. If you find this difficult to follow, don't worry—we shall return to it in the calculation details later on.

Calculation of a confidence interval

To derive the confidence interval for a proportion, we must first know the sampling distribution for a proportion. Consider a large population divided up into two types, A and B, in the proportions p and q respectively, so that $q = 1 - p$. Then the number of A's in the sample will follow a binomial distribution which has mean np and SD $\sqrt{(npq)}$. The *proportion* of type A is found by dividing the number of A's by n, and so the **sampling distribution of a proportion** has mean p and SE given by

$$\sigma_p = \sqrt{(pq/n)} \qquad (9.1)$$

Since the mean is p, a sample proportion is an unbiased estimator. The sampling distribution follows the binomial form but, provided that n is at least 15 or so (and larger if p is close to 0 or 1), the normal distribution is a good enough approximation. An allowance for continuity is necessary (this is explained later), but often it is too small to be important.

Suppose there are x A's in the sample of n. Then the observed proportion is x/n, and this must be used as a point estimate of the population proportion for the purpose of calculating the SE, using equation 9.1. The error involved is usually small since, if p is overestimated, q will be underestimated, and the product will turn out about right—as long as p is not too close to 0 or 1.

Initially, the formula for the confidence interval will be worked out for the simple case in which the sample is large enough for the continuity correction to be neglected. The observed proportion x/n is then a random observation from a normal population having mean p and SD σ_p. There will be therefore a 95% probability that x/n will be within $1.96\sigma_p$ of the value of p, with corresponding figures for any other percentage. This is equivalent to saying that, with 95% confidence, p will be within $1.96\,\sigma_p$ of x/n, i.e. that the 95% confidence interval for p is

$$p = x/n \pm 1.96\,\sigma_p \qquad (9.2)$$

The reason for the change of wording in the last sentence, from 'probability' to 'confidence', is given at the foot of page 98.

Example 1 213 drivers out of a random sample of 1000 failed a standard eye test. Find the 95% confidence limits for the proportion of all drivers who would fail the test. Here $x/n = 0.213$ and σ_p is estimated as $\sqrt{(0.213 \times 0.787/1000)} = 0.01295$. Hence the 95% limits are $0.213 \pm 1.96 \times 0.01295$, or 0.213 ± 0.0254. Notice that, if the actual value is the lower limit of 0.1876, the value of σ_p is 0.01235, and there is a corresponding difference for the upper limit. The resulting adjustments to the confidence limits are hardly large enough to need making; this is so in most cases, but care should again be taken if p is near 0 or 1.

Now consider the case when allowance for the continuity correction must be made. (There can be no general rule for this—it depends how accurately the limits are required to be known.) If x/n is greater than p, the value of x must be reduced by $\frac{1}{2}$, giving $(x-\frac{1}{2})/n-p < 1\cdot96\,\sigma_p$; if x/n is less than p, x must be increased by $\frac{1}{2}$, giving $p-(x+\frac{1}{2})/n < 1\cdot96\,\sigma_p$. (It is not easy to see which way the $\frac{1}{2}$ should go; for a significance test the reasoning is easier, as we shall see later, and clearly the correction must go the same way for both.)

These equations can be written

$$0 > (p-x/n) > (1\cdot96\,\sigma_p+1/2n),$$

and

$$0 < (p-x/n) < (1\cdot96\,\sigma_p+1/2n),$$

and then combined together to give the confidence limits for p:

$$p = x/n \pm (1\cdot96\,\sigma_p+1/2n) \tag{9.3}$$

Ignoring the correction thus gives a narrower interval than the true one. To give some idea of the magnitude of the correction, if $n = 100$ and $p = \frac{1}{2}$, σ_p is $0\cdot05$, so that a correction of $0\cdot005$ should be added to about $0\cdot1$, increasing the interval by 5%.

Example 2 A sample of 100 marbles, drawn from a large bag containing reds and blacks, yields 50 reds. Construct a 99% confidence interval for the proportion of reds in the bag. (This was the example used earlier.)

Using $p = \frac{1}{2}$, $\sigma_p = \sqrt{(\frac{1}{2} \times \frac{1}{2}/100)} = 0\cdot05$. For a 99% range, the limiting value of z, the standardised variate, is $2\cdot576$, and so the confidence interval is

$$p = 0\cdot5 \pm (2\cdot576 \times 0\cdot05 + 0\cdot005)$$

$$= 0\cdot5 \pm (0\cdot129 + 0\cdot005) = 0\cdot5 \pm 0\cdot134.$$

If $p = 0\cdot634$, the upper limit, a closer estimate for σ_p is $0\cdot0482$ instead of $0\cdot05$. This makes only a small change in the limits, to $0\cdot5 \pm 0\cdot1292$—not really worth considering. With suitable accuracy, the 99% confidence interval is therefore $0\cdot37$ to $0\cdot63$, as quoted earlier.

Now consider once more the alternative definition of a confidence interval: a random variable of a certain width (in graphical terms) and a known distribution, which has a probability such as 95% of containing the fixed but unknown value of the parameter. This is rather like trying to trap a mouse in a dark room by dropping an upturned bucket on to it; we cannot see the mouse, but the sounds it makes give us a rough indication of where to drop the bucket. The size of the bucket determines the margin of error which our point estimate of the correct position may have while still catching the mouse—the larger the bucket, the higher the probability. In the example above, we have an interval of 0·26, and the point estimate of its position is that the centre should be at 0·5. There is then a 99% probability that this interval will 'trap' the actual proportion. But, unfortunately, we can't take the bucket off to find out.

Calculation for a significance test

(i) A null hypothesis is proposed, usually on a basis of symmetry for problems of the biased-coin type or, more generally, by assuming that an observed effect which appears to be systematic is, in fact, due to chance. For example, if more patients appear to recover after treatment with a particular drug than without it, the null hypothesis would be that the drug had no effect and that the differences in the recovery rates were no more than the ordinary variations due to chance.

(ii) The probability is calculated that a result as far from the postulated one as that of the data could occur in a random experiment.

(iii) If this probability is over 5%, no conclusions can be drawn: the evidence tends to support the null hypothesis rather than disprove it, but the hypothesis is in no sense proved. If the probability is less than 5%, the data is significantly different from the hypothesis, and so the hypothesis is unlikely to be correct. The figure of 5% is fairly arbitrary, but is generally accepted as being the starting point of statistical significance. Conventionally, a figure between 5% and 1% is said to be *significant*, between 1% and 0·1% *very significant*, and less than 0·1% *highly significant*. The lower the percentage, the more significant the result.

Note, by the way, that statistical significance is not the same as practical importance. For very large samples (e.g. the ages of men married in England and in Scotland in a particular year) a very small absolute difference between the means—a month, perhaps—would be significant. But the fact thus established, that Englishmen (say) married at an earlier age than Scotsmen, would hardly be of practical importance. Conversely, small samples may not give a sufficiently sensitive test to establish a difference that is, in fact, quite important.

Example 3 A coin is spun 100 times and shows (a) 55, (b) 61, (c) 67, (d) 73 heads. Is the coin a fair one?

(i) The null hypothesis is that the coin is fair, with $p = \frac{1}{2}$.

(ii) We require the probability of a proportion at least as far from the mean of 0·5 as each observed proportion in turn. For 55 heads, the proportion is 0·55 on a discrete scale, but 0·545 on a continuous scale since, for 55 coins *or more*, the limit is the lower side of the 0·55 block on the discrete scale. (Compare the continuity correction made in the confidence interval formula.) Using equation 9.1, $\sigma_p = 0·05$, and so the limiting value of the standardised variate is 0·9. The corresponding figures for the other cases are 2·1, 3·3 and 4·5. The required probabilities are the areas of *both* tails of the distribution outside these values, and are 0·368, 0·0358, 0·000 96 and 0·000 003 4. The last figure here comes from Table 4.

(iii) (a) Results as far from the mean as 55 will happen with an unbiased coin about one in every three trials, and so no conclusions can be drawn.

(b) This is significant at the 5 % level, the actual figure being 3·6 % to two figure accuracy. So one could suggest that the coin was biased, at least if it were just an academic point. If one's reputation depended on it, it would be wise to be circumspect, because such a result will occur by chance with a fair coin about once in every 28 trials.

(c) This is highly significant, with a level of just less than 0·1 %. One could stake one's reputation with rather more confidence this time, and allege bias almost without reservation.

(d) Anything over 4 SD's can be counted as practical certainty for most purposes. If a fair coin were spun continuously at the rate of 100 throws every five minutes, then, in **each** 100, a result as far from the mean as 73 would occur only about **once** every three years.

Example 4 A large bag of marbles, which is believed to contain a proportion 0·37 of reds, yields 50 reds in a sample of 100. Is this in accordance with the proportion given? (This is the illustration previously used, and the figure of 0·37 has been chosen to facilitate comparison with the confidence interval calculation.)

The SE is $\sqrt{(0·37 \times 0·63/100)} = 0·0482$. Hence the standardised variate corresponding to 50 is $(0·495–0·37)/0·0482 = 2·59$. The probability of a result as far from the postulated mean as the observed figure is therefore 2×0.0048 or about 0.01, and so the result is significant at the 1% level. This agrees with the previous calculation, in which 0.37 was found to be the lower limit of the 99% confidence interval on the same data.

One-tail and two-tail tests

In example 3(b), the significance level was given as 3.6% for the probability that a random result would be in one of the two tails of the distribution outside the given value of z. Two-tail tests, of which this is an example, are the more common type, but in certain circumstances a one-tail test is used. Consider a variation of the biased-coin experiment, in which the coin is alleged to be biased towards the head, and the allegation is tested by spinning the coin 100 times. Sixty-one heads appear. The bias is confirmed, but at what significance level? Here, any number of heads less than fifty would not have indicated bias *of the type alleged*, and so this tail is irrelevant for the test being made. The significance level is therefore the probability corresponding to one tail, which is 1.8%. As a general rule, a one-tail test is only used if there is a suggestion of bias *in a specific sense* before the result of the experiment is known.

We have now covered the general principles of significance tests and confidence intervals, using a proportion as an example of a parameter to be estimated. We shall now do the corresponding methods for the estimation of a population mean from sample data: most of the formulae are almost identical, but the estimation of the population SD is not so easy.

Confidence intervals for a population mean—large samples

It follows at once from equation 8.2 that a confidence interval for the population mean is given by

$$\mu = m \pm z\sigma/\sqrt{n}, \qquad (9.4)$$

where μ and m are the population and sample means respectively, σ is the population SD, and n the number in the sample.

However, an approximation is involved on account of the inter-relationship between the mean and SD of the sample, which is negligible only if the sample is large. Small samples are considered in the next section, when it will also become clear what size may be considered 'large'.

Although, as we have seen, m is an unbiased estimator of μ, s is not an unbiased estimator of σ—it consistently tends to underestimate it. The sample SD is calculated from the sum of squared deviations from its own mean, not (as must be done for an unbiased estimate) from the population mean. Thus, since the sum of squared deviations is a minimum when the deviations are measured from the mean of the actual variates, s is an underestimate of σ. An unbiased estimator can be set up as follows. The population SD is defined by the formula

$$N\sigma^2 = \sum_1^N (x-\mu)^2.$$

But only n measurements are available, and so the best (and unbiased) estimate of σ is given by

$$n\hat{\sigma}^2 = \sum_1^n (x-\mu)^2,$$

where the 'hat' over the σ is used to denote an estimated value. For the sample, the SD may be calculated as follows, using the population mean μ as a working mean:

$$ns^2 = \Sigma(x-\mu)^2 - n(m-\mu)^2.$$

Combining the last two equations together, and cancelling n's,

$$\hat{\sigma}^2 = s^2 + (m-\mu)^2.$$

It remains to give an unbiased estimate of $(m-\mu)^2$, and such an estimate is the mean of the probability distribution of this quantity. But the SE of the mean is defined as the square root of this very thing, and hence

$$(1/n)\Sigma(m-\mu)^2 = \sigma_n^2 = \sigma^2/n.$$

Thus $\hat{\sigma}^2 = s^2 + \sigma^2/n$, and, finally, using $\hat{\sigma}$ for σ,

$$\hat{\sigma}^2 = \frac{n}{n-1}s^2. \tag{9.5}$$

It also follows that

$$\hat{\sigma}^2 = \frac{\Sigma d^2}{n-1} \tag{9.6}$$

if $s^2 = (1/n)\Sigma d^2$; that is, for a population estimate of the SD rather than the sample value, the usual form is modified by substituting $n-1$ for n.

We now return to the confidence interval for the mean. Using equation 9.4, the 95% confidence interval is

$$\mu = m \pm 1\cdot96\frac{s}{\sqrt{n}}\bigg/\!\left(\sqrt{\frac{n}{n-1}}\right)$$

$$\text{i.e. } \mu = m \pm 1\cdot96\frac{s}{\sqrt{(n-1)}}. \tag{9.7}$$

Almost invariably $\hat{\sigma}$ is calculated directly from the sample data, using divisor $n-1$ as in equation 9.6, and this will be done from now on unless otherwise indicated. Thus the most convenient form for the confidence interval is

$$\mu = m \pm z\hat{\sigma}/\sqrt{n} \tag{9.8}$$

where z is the appropriate normal variate.

Example 5 A sample of 100 taken from a large population has mean 57 and gives an estimated population SD of 4. What are the 99% confidence limits for the population mean? (A sample of 100 is large enough for the normal distribution to be used, as will be seen later.)

Using equation 9.8, $\mu = 57 \pm 2\cdot58 \times 4/\sqrt{100} = 57 \pm 1\cdot03$.

Example 6 The Ruritanian Government wishes to fix a standard height for a doorway, high enough for 99% of the male population to pass through without bending, allowing 4 cm for shoe thickness and clearance at the top. A sample of 300 adult males has mean height 168 cm and gives an estimated SD of the population of 8·5 cm. What height should be fixed?

It is necessary first to fix the upper confidence limit for the population mean—the confidence level is not specified, and a 95% one is appropriate. This gives a limit of $168 + 1\cdot96 \times 8\cdot5/\sqrt{300} = 168\cdot96$ cm.

For a height which cuts off a 1% tail, the standardised normal variate is 2·33, i.e. the height required is $2\cdot33 \times 8\cdot5$ cm above the estimated highest value for the mean, plus the margin of 4 cm. This is $168\cdot96 + 2\cdot33 \times 8\cdot5 + 4 = 192\cdot7$ cm, or say 193 cm.

Confidence intervals for a population mean—small samples

For this it is necessary to assume a normal population, although the results may be used with fair accuracy if the departure from normal is not too great.

Consider a sample of n having mean m, which gives an estimated population SD of $\hat{\sigma}$. This is assumed to be drawn from a normal population of which neither mean nor SD is accurately known. If it had been drawn from a population of known SD σ, we should require the value of the normal variate defined by

$$z = (m - \mu)/(\sigma/\sqrt{n}) = \frac{(m - \mu)\sqrt{n}}{\sigma}.$$

For a small sample, the method is to consider the distribution of a similar quantity, given by

$$t = \frac{(m - \mu)\sqrt{n}}{\hat{\sigma}}.$$

This is not normally distributed, because the variable $\hat{\sigma}$ has to be used in place of the fixed σ. (By using equation 9.5, an alternative form can be derived for s instead of $\hat{\sigma}$, but this is less useful because, in practice, $\hat{\sigma}$ is almost always calculated directly from the sample data.)

The actual distribution was derived by W. S. Gosset, who wrote under the pen-name of 'Student'. As with the normal distribution, the mathematics need not concern us, and we require only the tabulated values of Gosset's t. Values for some typical probability levels are given in Table 7. The probabilities are those for a two-tailed test.

The calculation is exactly as before, except that the new statistic t replaces the previous z. Equation 9.8 thus becomes

$$\mu = m \pm t\hat{\sigma}/\sqrt{n} \tag{9.9}$$

The quantity for which the probabilities are tabulated is not the number in the sample but the number of what are called **degrees of freedom** in the sample. This is usually abbreviated to d.f., and denoted by v, the Greek nu. In the present case, $v = n - 1$. The reason for the name is that, for a sample of given size n and given mean, the variates are not all independent: after $n - 1$ of them are fixed, the remaining one has to take a particular value in order to give the known value for the mean. The term is discussed in more detail in chapter 13.

We can now form some estimate of what size sample can be counted as 'large' when deciding whether to use the normal or t-distribution. The values shown under $v = \infty$ are the percentage points of the normal curve and, for $v = 120$, the values are not very different. Anything over about 100 therefore counts as a large sample for most purposes, and often the limit is taken as 30 or 25. For a sample of 25 ($v = 24$), the difference might be enough to matter—2·80 compared with 2·58 for a 1% two-tail probability. For very small samples, the differences become much larger.

Example 7 A biscuit-packing machine produces packets of nominal mass 250 g. A sample of five has masses 252, 257, 255, 251 and 258 g. What are the 95% confidence limits for the mean of the output?

The sample mean is 254·6 g. The estimated population variance is $\frac{1}{4}(2·6^2 + 2·4^2 + 0·4^2 + 3·6^2 + 3·4^2) = 9·3$, and hence $\hat{\sigma} = \sqrt{9·3} = 3·05$. The value of t for a 95% confidence limit is found in Table 7 under 4 d.f., that is, $t = 2·78$. Hence the required limits are $254·6 \pm (2·78 \times 3·05)/\sqrt{5}$, or $254·6 \pm 3·8$ g.

Example 8 A series of 10 measurements of the acceleration due to gravity at a certain point gave a mean of 9·8065 m/s², and the sum of the squared deviations from this value was $7·62 \times 10^{-5}$ m²/s⁴. Find the 99% confidence limits for the correct value.

The estimated population SD is $\sqrt{(\frac{1}{9} \times 7·62 \times 10^{-5})} = 2·91 \times 10^{-3}$ m/s². For 99% confidence and 9 d.f., $t = 3·25$. Hence the required limits are

$$9·8065 \pm 3·25 \times 2·91 \times 10^{-3}/\sqrt{10} = 9·8065 \pm 0·0030 \text{ m/s}^2.$$

Significance tests for a sample mean

These follow the methods already outlined, using the SE of the mean and either the normal or the t-distribution as appropriate. The null hypothesis is that the population mean takes a particular value, and the difference between that and the sample mean may or may not be significant.

Example 9 A machine produces components with a nominal dimension of 8·000 cm. A sample of 4 has mean 8·007 cm and gives an estimate of 0·004 cm for the population SD. Is there evidence that the mean of the output is not 8·000 cm?

Here

$$t = \frac{(m - \mu)\sqrt{n}}{\hat{\sigma}} = \frac{0·007\sqrt{4}}{0·004} = 3·5.$$

Using the t-distribution for 3 d.f., this is significant at the 5% level. The mean of the output is probably set too high.

Example 10 Four marksmen score 85, 88, 81 and 92. With a different type of sight, they score (in the same order) 89, 87, 87 and 95. Can the general improvement be attributed to chance?

The improvements in the individual scores are 4, -1, 6 and 3. We have to decide whether these could reasonably have come from a normal population (which is a reasonable assumption for the distribution of random differences in scores) having a mean of zero. The sample mean is 3, and the estimated population variance is $\frac{1}{3}(1^2 + 4^2 + 3^2 + 0^2) = 8\frac{2}{3}$. Hence $t = 3\sqrt{4}/\sqrt{8\frac{2}{3}} = 2\cdot04$. For 3 d.f. this is not significant even at the 10% level, and so more tests are required before it can be said that the new sight is an improvement on the old.

This type of test is known as a **paired sample test**. There is a single random variable involved, which here is the difference between a marksman's two scores. If the scores could not be paired off in this way (for example if a group of eight marksmen had been divided randomly, four using one sight and four the other), a different test would have to be used, and this is discussed in chapter 10.

False conclusions from significance tests

There are two ways in which a significance test can mislead. They are not mistakes in the $2 + 2 = 5$ sense, but nevertheless they are ways in which a wrong conclusion is reached. First, we can decide that the null hypothesis is wrong when in fact it is right, and the significance level is a measure of the probability of doing so: this is known conventionally as a Type I error. We can also decide that the evidence does not indicate that the null hypothesis is wrong when in fact it *is* wrong: this is known as a Type II error, and is very much more common. If a coin is biased with $p = \frac{2}{3}$, for example, only about half the trials of 100 spins will indicate bias at a significance level of $0\cdot1\%$. (The trials are, of course, considered separately—the cumulative effect of several trials would soon provide any level of significance required.)

It is, of course, desirable to arrange the conditions of the test in such a way that the chance of a Type II error is as low as possible; the probability of avoiding such an error (which should be as high as possible) is known as the **power** of the test. The most obvious way of increasing the power of the test is to increase the sample size.

Procedure in a court of law is a useful comparison. Since a man is assumed innocent until proved guilty, the null hypothesis is that he is innocent. A Type I error is to decide wrongly that the null hypothesis is untrue, that is, to declare an innocent man guilty. This is a serious matter and, within the limits of human fallibility, must not be allowed to happen. A Type II error is to let a guilty man go free and, although this is regrettable, it is not as serious as a Type I error, and happens much more often.

The three methods compared

A point estimate has the merit of simplicity but, because there is no indication of how close it is to the truth, it is at best limited in usefulness, and at worst highly misleading.

A significance test is the next easiest to apply, and is a powerful and much-used test. Its main limitation is the nature of the conclusions which can be drawn: a theory can be discounted but, except in a very limited sense, can never be confirmed. Further, when a theory is discounted, no quantitative conclusions can be drawn as to whether it is slightly inaccurate or totally wrong. This form of conclusion is not as easy for a layman to understand as that obtained with a confidence interval, but it is a method of more general application. For example, a six-sided die can be tested for bias, using a significance test of the type discussed in chapter 13, although there is no single measure for which to give a confidence interval. Another example is the testing of the effectiveness of a drug. A significance test is also the best way of comparing the statistics of two samples, and this is discussed in the next chapter.

A confidence interval is a little more difficult to calculate, but the results are generally more useful as well as being easier to interpret. Where the quantity being tested has no obvious theoretical value, such as $p = \frac{1}{2}$ for a coin, a confidence interval will commonly be used. Its main limitation is that there must be some quantity for which to give an interval, as mentioned above.

Both significance tests and confidence intervals are based on the assumption that the test statistic follows the normal or t-distribution.

A note on Bayesian methods

Suppose we have a number of bags, one-quarter of which contain two red marbles, the remainder containing one red and one black marble. We take a bag at random, and draw one marble, which proves to be red. What is the probability that the other marble in that bag is red? This is not a difficult problem, and can be answered as follows. The possible results of the experiment are: choose red-only bag, draw red (probability $\frac{1}{4}$); choose mixed bag, draw red ($\frac{3}{4} \times \frac{1}{2} = \frac{3}{8}$); choose mixed bag, draw black ($\frac{3}{4} \times \frac{1}{2} = \frac{3}{8}$). The relative expected frequencies of drawing a red, from an all-red bag and from a mixed bag, are therefore in the ratio $\frac{1}{4}$ to $\frac{3}{8}$, and so the probability that a red marble comes from a red-only bag is $\frac{2}{5}$.

Now repeat the experiment with one difference—that the proportion of red-only bags is unknown. (This simple illustration represents, mathematically, quite a common practical situation.) The problem is now insoluble under the usual terms of reference, and the best we can do is to assume that the unknown probability of choosing a red-only bag is $\frac{1}{2}$. This is not a probability in either of the senses previously defined: it is a numerical interpretation of a degree of belief. There are two possibilities—red-only and mixed—and, since we have no evidence which makes one more likely than the other, we assume that they are equally probable. The probabilities of the two ways of choosing a red are therefore taken to be $\frac{1}{2}$ from a red-only bag and $\frac{1}{2} \times \frac{1}{2} = \frac{1}{4}$ from a mixed one; and so the probability that it is a red-only bag is $\frac{2}{3}$.

This method was first used by Thomas Bayes (1702–1761), and the assumption made above is an example of what is known as Bayes' Postulate. He suggested that, when probabilities were unknown before an experiment was carried out, they were to be assumed equal.

The extent to which this is a fair assumption has been for many years a highly contentious matter, and further consideration of it is out of place here. Bayesian methods are undoubtedly very useful in providing ways of evaluating confidence limits where frequency methods cannot do so, but they will not be used in this book.

Exercises

Confidence intervals for a proportion

1 In the drawing pin experiment described in chapter 4, 794 heads appeared in 2000 throws. What are the 95% confidence limits for the probability of a head?

2 Of 7500 live births in a city in one year, 3879 were boys. Derive the 99% confidence limits for the probability than an unborn child will be a boy.

3 A die was thrown 1000 times and 215 sixes were noted. Show that it is reasonable to conclude that the die was biased, by calculating the 95% confidence limits for the probability of a six.

4 Before a by-election, in which there are two candidates A and B, enquiries are made of 400 voters, chosen at random, and it is found

that 208 of them intend to vote for A. Give 95% confidence limits for the percentage of voters favourable to A at the time of the inquiry. If, in fact, 55% of the voters were in favour of B, what is the probability that a random sample of 400 voters will contain at least as many in favour of A as there are in favour of B?   (AEB)

5 If needles of length a are thrown randomly on to parallel lines distance a apart, the probability that a needle will fall across a line is $2/\pi$. In 10 000 throws, 6443 needles fell across a line; derive 95% confidence limits for the value of π.

6 How large a sample should be taken in a political opinion poll if there are two parties, fairly evenly matched, and the two proportions are required to be within 1% of the correct figures, with 95% confidence?

Significance tests for a proportion

7 In the drawing pin experiment described in chapter 4, the numbers of heads after each 40 throws up to 320 were as shown in the table below:

Throws:	40	80	120	160	200	240	280	320
Heads:	17	34	52	65	85	97	114	125

If it was suggested that the probability of a head was in fact $\frac{1}{2}$, what conclusions could be stated at each stage?

8 A roulette wheel is observed to show a zero ($p = 1/37$) 322 times in 10 000 spins. Is there any reason to doubt its fairness?

9 In a test of the randomness of the last digits of some telephone numbers, 82 ones were counted in 1000 numbers. Is there any evidence of non-randomness?

10 A pack of cards contains five different patterns in equal numbers. A card is drawn at random; it is then replaced, the cards are shuffled, and another card is drawn at random. A man claims that he can identify the pattern drawn, more often than once in five, without seeing the card. In 100 drawings, how often at least must he correctly identify the pattern in order to justify his claim at the 5% level of significance?   (AEB)

11 Over a long run, the proportion of defectives in a manufacturing process is constant at 8%. A sample of 50 has 11 defectives. How strong is the evidence that the proportion of defectives has altered?

12 In the Republic of Dexteria, the general incidence of left-handedness is 7%, but in the town of Siniston the proportion is 8% of its 5000 inhabitants. Can the difference be attributed to chance?

13 A population consists of two types in proportions p and q, and two alternative values are suggested for p: that $p = \frac{2}{3}$ and that $p = \frac{3}{4}$. It is required to take a sample to test which is correct. Determine how large a sample is necessary so that, whatever the value of the observed proportion, it must be significant at the 5% level when compared with at least one of the hypotheses.

Confidence intervals for a population mean—large samples

14 A sample of 100 has mean 57·4 and SD (divisor $n-1$) of 4·6. What are the 95% confidence limits for the population mean?

15 A sample of 75 has mean 38·6 and SD (divisor $n-1$) of 2·2. What are the 99% confidence limits for the population mean?

16 A sample of 80 has mean 71·2, and the sum of the squared deviations from this figure is 487. What are the 95% confidence limits for the population mean?

17 For a sample of 120, the sum of the observations was 4792 and the sum of the squared observations was 192 708. What are the 99% confidence limits for the population mean? (*Note*: the use of four-figure tables or a slide rule will lead to difficulties in this question.)

18 A sample of 50 ball-bearings has mean diameter 0·5002 cm and SD (divisor $n-1$) 0·000 17 cm. Find: (a) the 95% confidence limits for the mean of the population; (b) the rejection rate if the tolerances are $0·5000 \pm 0·0005$ cm and the population mean is 0·5002 cm; (c) the rejection rate with the same tolerances if the population mean is at the upper 95% confidence limit just found.

19 A sample of 35 has mean 89·6 and SD (divisor $n-1$) 11·3. Below what figure may 90% of the distribution be expected to lie, with 95% confidence?

20 How large a sample should be taken from a normal distribution in order to establish the value of the population mean to within one-fifth of the size of the population SD, with 95% confidence?

21 How large a sample should be taken from a normal distribution in order to establish the value of the population mean to within one-tenth of the size of the population SD, with 99% confidence?

22 A factory produces washers of mean thickness 3 mm with a SD of 0·2 mm. Each day, as a routine check, the total thickness of a random

sample of 100 washers is measured, and the mean calculated and recorded. Every 30 days, these means are assembled into a frequency table. What would you expect the mean and SD of this table to be? On one day, the mean thickness of the sample of washers is 3·036 mm. Is this significantly different from the expected value?   (OL)

23 A machine is supposed to produce components to a nominal dimension of 5·000 cm. A random sample of 100 components produced by the machine has a mean of 5·008 cm and SD (divisor $n-1$) of 0·036 cm. Estimate the SE of the mean, and obtain a 95 % confidence interval for the mean of the whole output. Test whether, on the evidence of the sample, the mean of the whole population differs from 5·000 cm (a) at the 5 % level of significance, (b) at the 2 % level of significance.   (JMB)

24 An anthropologist wishes to know whether the adult males on an island are significantly different in height from those on the mainland. Figures for the mainland population are known to be 171·2 cm for the mean height and 8·9 cm for the SD; a sample of 113 on the island has mean height 169·8 cm. What conclusions should be drawn?

25 An automatic lathe produces components with a nominal diameter of 6·5 cm and is known to operate with a SD of 0·005 cm. A sample of 50 has mean diameter 6·497 cm. Is there evidence that the machine requires adjustment?

Paired sample tests—large samples

26 In order to test the effect of a rust-proofing process on the strength of a metal, 100 pieces are cut in two, and one of each pair only is treated. In suitable units the average excess of strength of the treated pieces over the untreated pieces is 2·1, and the SD of the excess is 10. Stating all assumptions and giving full explanations, carry out a significance test at the 5 % level to see whether the treatment affects the strength.   (JMB)

27 Two versions of an 'intelligence test', designed to be of the same standard of difficulty, are compared by giving each one in turn to 200 people. The difference between the mean scores is 3·6, and the SD of the differences is 14·7. Can it reasonably be assumed that the two standards are the same?

28 In a test of the comparative standards of two examination boards which followed the same syllabus, 420 candidates sat the examinations of both. Each board graded its results from 1 to 9, and the differences

in grades had a mean of 0·135 and a SD of 1·62. Can the standards reasonably be considered to be the same?

Small sample tests

29 A sample of 10 was drawn in a manner suggested to be random from a population known to have a mean of 100, and gave the following figures: 87, 98, 91, 100, 97, 105, 102, 88, 90, 97. Is there any reason to doubt the randomness?

30 A sample of 10 gave the following figures: 23·5, 24·8, 24·0, 23·4, 24·6, 25·1, 23·6, 24·2, 24·2, 24·6. Find the 95% and the 99% confidence limits for the population mean, assuming a normal distribution.

31 A sample of five items has masses 13·7, 12·9, 13·4, 12·8 and 13·2 g. What are the 95% confidence limits for the mean of the population?

32 Eight fish of a certain species are measured, and their lengths are found to be 20·6, 21·2, 20·4, 22·2, 21·3, 20·2, 20·3 and 22·5 cm respectively. Find the 95% confidence limits for the mean length of a fish of this species, assuming a normal distribution, and estimate the length which will be exceeded by only one fish in a thousand. (AEB)

33 Eight golfers take 77, 83, 81, 76, 88, 80, 74 and 83 strokes respectively to go round an 18-hole course, using identical sets of clubs. They are then asked to try out sets of clubs of a new type, and (in the same order) take 74, 73, 84, 78, 79, 79, 72 and 83 strokes over the same course. Can it be said that the new type is an improvement on the old? Comment on the assumptions you have made.

34 A mass-produced component should have a mean diameter of 2·500 cm. A sample of 15 has mean diameter 2·4963 cm and the sum of squared deviations from this is 0·0806 cm². Is there evidence that the mean is incorrectly set?

35 Eight typists were given electric typewriters instead of their previous machines, and the increases in the numbers of finished quarto pages per day were found to be 8, 12, −3, 4, 5, 13, 0 and 11 respectively. Do these figures indicate that the typists in general can work faster with electric typewriters? State the assumptions you make. (AEB)

36 Ten measurements of a physical constant gave a mean of 723·77 units, and the sum of the squared deviations from this figure was 0·9261 squared units. What are the 95% confidence limits for the actual value of the constant?

Chapter 10

More significance tests

The significance of a difference between two proportions

Suppose we have a proportion p_1 in a sample of n_1, and p_2 in a second sample of n_2. Then we can test whether the difference between p_1 and p_2 is significant, on the null hypothesis that the two samples are drawn from a single binomial distribution.

It is necessary first to estimate the value of p for this distribution, and to do this we combine the samples together. This gives

$$p = \frac{n_1 p_1 + n_2 p_2}{n_1 + n_2}. \qquad (10.1)$$

The distributions of p_1 and p_2 are approximately normal (as in the single-proportion case discussed in chapter 9), each with mean p and with variances pq/n_1 and pq/n_2, where $q = 1 - p$. In chapter 8, we derived the fact that the variance of the difference between two variates is equal to the sum of the two variances; the variance of the difference between two proportions is therefore $pq/n_1 + pq/n_2$, and the **SE of the difference between two proportions is**

$$\sigma_d = \sqrt{\left\{ pq \left(\frac{1}{n_1} + \frac{1}{n_2} \right) \right\}} \qquad (10.2)$$

In order to find the form of the distribution, it is necessary to know that the difference between two normal variates is itself a normal variate. This should be thought reasonable enough by anyone who has followed the theory of distributions thus far, though the actual proof is too difficult to tackle here. The difference between the proportions is therefore normally distributed.

Example 1 In a sample of 500 Irishmen, 17% had red hair; in a sample of 1500 Englishmen, 15% had. Is the difference significant? The overall value of p is 0·155, and hence the SE of the difference is

$$\sqrt{\left\{0{\cdot}155 \times 0{\cdot}845 \times \left(\frac{1}{500}+\frac{1}{1500}\right)\right\}} = 0{\cdot}0187.$$

The observed difference of 0·02 is only just over 1 SE, and not significant.

The significance of a difference between two means

This is one of the most useful significance tests. One typical problem is this: the average height of plants grown with one fertiliser is more than the average height with another type—is this conclusive evidence that the first type is better? The method used depends on three distinct conditions, and it is necessary to be very clear about which of these obtain before starting the actual calculation. The conditions are:

1 Are the samples large? This is to determine whether the sample means follow a normal or a t-distribution: the total in the two samples should be at least 25 before the normal can be used as a rough approximation, and about 100 before it can be used more or less exactly.
2 Are the samples drawn from normal populations?
3 Is there a significant difference between the two sample SD's? If so, the null hypothesis must be that the two samples are drawn from different populations having the same mean; if not, it is more satisfactory to assume that they are drawn from a single population.

Differences between two means—large samples

The test is a straightforward one to apply, but the formula used for the SE depends on whether one or two populations are assumed. Consider first the case of two populations.

Since the samples are large, the sample means are normally distributed whatever the original distributions. Suppose we have samples of n_1 and n_2, sample means m_1 and m_2, and estimates of the population variances (derived from the samples) of $\hat{\sigma}_1^2$ and $\hat{\sigma}_2^2$. The variances of the sample means are then $\hat{\sigma}_1^2/n_1$ and $\hat{\sigma}_2^2/n_2$, and the variance of the difference equals $\hat{\sigma}_1^2/n_1 + \hat{\sigma}_2^2/n_2$. The normally distributed difference has therefore to be compared with the **SE of the difference**, which is

$$\sigma_d = \sqrt{\left|\left(\frac{\hat{\sigma}_1^2}{n_1}+\frac{\hat{\sigma}_2^2}{n_2}\right)\right|} \tag{10.3}$$

Example 2 Tests are carried out to see if a fabrication process is more suited to men or women. A sample of 150 men do the job with a mean time of 33 minutes and SD of 4 minutes; 100 women have a mean of 31 minutes and SD of 6 minutes. Are the women significantly faster? The SE of the difference is

$$\sqrt{\left(\frac{4^2}{150}+\frac{6^2}{100}\right)} = 0.683 \text{ minutes,}$$

and so the observed difference of two minutes is very significant, at a level of about 0.35%.

Taken on its own, this test has severe limitations, because it takes no account of the difference between the SD's—and this may be just as important as the difference between the means. In example 2 it can easily be shown that the slowest 15% of the women are slower than the corresponding proportion of the men. In practice, this might easily outweigh the advantage of the women judged on the mean. If the figures appear to indicate that the difference between the SD's might be important, a preliminary test must be made to find whether this is so.

The test statistic is the **variance ratio**, denoted by F, which is the ratio of the larger of the variances (defined with divisor $n-1$) to the smaller. The distribution of F has been tabulated for different numbers of degrees of freedom (equal here to $n-1$) for the two samples, and a table of the 5% and 1% points is given at the end of the book as Table 9. v_1 is the number of d.f. for the sample with the larger variance.

In example 2, the ratio is 2.25, which is well above the 1% limit—it is, in fact, significant at the 0.1% level, and so, as is fairly obvious, the variability of the women's times is significantly greater.

Where both the variance ratio and the difference between the means are significant, the results of the two tests must be considered together, and the conclusion will depend on the relative importance which is attached to each. We now go on to consider the case where the variance ratio (either by using the F-test or by inspection) is not significant.

In this case we can assume, as a null hypothesis, that the two samples are drawn from a single population. This is much more satisfactory logically than two populations, because we assume now that there is no significant difference either in mean or SD between the two samples. Previously we had to consider different SD's, which makes the assumption that the means are the same inherently less plausible. In practice, a single population is also the more common case.

The best estimate for the SD of the single population must be found from the two samples together, as follows. The derivation follows that of equation 9.5, which should be consulted for fuller explanations. Suffixes 1 and 2 indicate the samples.

The best estimate is given by $(n_1 + n_2)\hat{\sigma}^2 = \Sigma(x - \mu)^2$, where the summation is over both samples in turn. For the first sample, $n_1 s_1^2 = \Sigma(x - \mu)^2 - n_1(m_1 - \mu)^2 = \Sigma(x - \mu)^2 - \hat{\sigma}^2$. Repeating for the second sample and combining the two, $n_1 s_1^2 + n_2 s_2^2 = \Sigma(x - \mu)^2 - 2\hat{\sigma}^2$, where the summation is now over both samples. Finally, this equation is combined with the first one to give $\hat{\sigma}^2$:

$$\hat{\sigma}^2 = \frac{n_1 s_1^2 + n_2 s_2^2}{n_1 + n_2 - 2}. \tag{10.4}$$

Alternatively, since $(n - 1)\sigma^2 = ns^2$ (eqn 9.5), the equation can be written

$$\hat{\sigma}^2 = \frac{(n_1 - 1)\hat{\sigma}_1^2 + (n_2 - 1)\hat{\sigma}_2^2}{(n_1 - 1) + (n_2 - 1)}. \tag{10.5}$$

The estimated population variance is thus the weighted mean of the two sample estimates, allowing a weighting $n - 1$ for each. This is a more convenient form, since the sample SD's are almost invariably calculated using divisor $n - 1$, as noted in the last chapter.

We now have two large samples whose means are normally distributed with variances $\hat{\sigma}^2/n_1$ and $\hat{\sigma}^2/n_2$. As before, the difference is also normally distributed, but this time the variance is $\hat{\sigma}^2/n_1 + \hat{\sigma}^2/n_2$, which gives

$$\sigma_d = \hat{\sigma}\sqrt{\left(\frac{1}{n_1} + \frac{1}{n_2}\right)} \tag{10.6}$$

This should not be confused with the form previously given as equation 10.3, although the two are obviously identical when $\hat{\sigma}_1 = \hat{\sigma}_2$. To illustrate the difference between them, consider two samples of 25 and 125, with SD's of 9 and 7. This gives a variance ratio which is just significant at the 5% level. Using equation 10.3, the value of σ_d is 1·90, and using 10.6 it is 1·61. Differences tend to be greater as sample size increases.

Example 3 Two versions of an intelligence test, designed to have the same standard of difficulty, were tried out as follows. A sample of 400 people was divided randomly, and one test given to each group of 200. The mean scores were 54·1 and 56·3, with SD's of 8·7 and 8·5. Are the standards of the two tests the same?

Here the difference between the SD's is small, and one does not need an F-test to see that it is not significant. A single population is therefore assumed, with SD 8·6, using equation 10.5. The SE of the difference is thus 0·86, and the observed difference of 2·2 is significant at about the 1% level. The standards of the two tests cannot be assumed to be the same.

Differences between two means—small samples

As with the significance of a single mean, the change to small samples involves two alterations to the method. First, it only applies exactly to samples drawn from normal populations, though if the departure from the normal is not too great, errors are not usually serious. Second, the difference between the sample means is no longer normally distributed.

Consider first the case where two small samples are assumed to be drawn from the same population, an F-test being used, if necessary, to check the validiaty of this. Following the method of the last section, the test statistic is the ratio between the difference of the two means and the SE defined by

$$\sigma_d = \hat{\sigma} \sqrt{\left(\frac{1}{n_1} + \frac{1}{n_2}\right)}.$$

This would be a standardised normal variate if σ could be used in place of $\hat{\sigma}$, but, since an estimate is used, the ratio follows the t-distribution, as in the last chapter. The number of degrees of freedom is $n_1 + n_2 - 2$, as used in calculating $\hat{\sigma}$. The significance test now continues as before.

Example 4 A new type of cattle feed was tested, and the results were as follows. For a sample of 10 calves, the average increase over 14 days was 13·7 kg, with a sum of squared deviations of 38·0 kg^2; for the control sample of 10 similar calves, the figures were 12·2 and 31·2. Is the new type better?

First, an F-test. The ratio of the variances is 38·0 : 31·2, which is 1·22, well below the significant figure of 3·2. The population variance is therefore taken to be $\frac{1}{18}(38·0 + 31·2) = 3·85$, and so $\sigma_d = \sqrt{\{3·85(\frac{1}{10} + \frac{1}{10})\}} = 0·878$. Hence $t = 1·5/0·878 = 1·71$, well below the 5% significance level for 18 d.f., which is 2·10. The new feed needs further testing.

Finally, we must see what happens when two small samples have variances which are significantly different. The obvious test statistic is

$$(m_1 - m_2) / \sqrt{\left(\frac{\hat{\sigma}_1^2}{n_1} + \frac{\hat{\sigma}_2^2}{n_2}\right)}.$$

But this does not follow a t-distribution, because two different population estimates are used, and so we have a different set of conditions from the one above. A new distribution is therefore required, and a search for it would take us far beyond the limitations of an elementary textbook.

Of the eight possible cases allowed by the three conditions first mentioned, we can therefore cope with all but three: small samples from a single non-normal distribution, and small samples from two populations whether normally distributed or not. Fortunately, a single population is the more common case and, indeed, unless it is very far from normal, no serious difficulties should be encountered.

Exercises

The difference between two proportions

1 Two firms of opinion poll organisers conducted surveys in a constituency where two parties were contesting an election. Ignoring the 'don't knows', the first survey showed 52 % for party A and 48 % for B in a sample of 800 voters; the second showed 47 % for A and 53 % for B in 500. Is there any reason to doubt the validity of one of the surveys?

2 Repeat the calculations of question 1 assuming that the samples were each 2000 in number.

3 A sample survey showed that, of 4000 random voters, 2273 were Republicans, and of these 1280 were men; of the 1727 Democrats, 822 were men. Are the proportions of men and women who support the two parties significantly different?

4 Identical advertisements were placed in two weekly journals. The first had a circulation of 47 000 and produced 385 enquiries; the second had a circulation of 34 000 and produced 312 enquiries. Is the response significantly different?

5 A manufacturer of two types of television set found that, for one model, 89 sets out of 5370 in one year's output required a new tube under guarantee; for the other, 135 out of 7160 required new tubes in the same period. Is there any evidence that the tube of one model is more susceptible to failure than the other?

6 In the drawing pin experiment described in chapter 4, the first 1500 drawing pins produced 581 heads, and the next 500 produced 213 heads. Is there any evidence that the conditions of the experiment had changed?

The difference between two means—large samples

(A variance ratio test is not required except in question 12.)

7 Two samples, each of 100, have means 57·2 and 58·9, and SD's (divisor $n-1$) 6·3 and 5·9, respectively. Is there a significant difference between the means?

8 Samples of 150 and 250 have means 81·6 and 82·9, and SD's (divisor $n-1$) 4·3 and 4·9, respectively. Is there a significant difference between the means?

9 Samples of 70 and 90 have means 35·1 and 35·8, and SD's (divisor $n-1$) 2·9 and 3·2, respectively. Is there a significant difference between the means?

10 Two samples of 80 and 120 have means 50·7 and 51·3, and the sums of squared deviations from these figures are 256·3 and 370·5 respectively. Is there a significant difference between the means?

11 A large square field was divided into 400 small squares, and alternate squares were given two different fertilisers in equal quantities. The crop was sown uniformly, and the yields from each square measured. For the first type, the mean yield was 26·2 kg and the sum of squared deviations from this figure was 1783 kg^2; for the second type the figures were 26·6 kg and 1960 kg^2. Is there a significant difference between the yields?

12 Two intelligence tests are given to groups of 300 people, who are chosen by random selection from a group of 600. One sample has mean 102·6 and SD 16·2, and for the other the figures are 98·5 and 12·1. Comment on the differences. (It may also be of interest to compare the merits and disadvantages of this test with the paired sample test—see exercise 27 of chapter 9.)

The difference between two means—small samples

(Variance ratio test required.)

13 Two different experiments were designed to measure the same physical constant, and each was carried out ten times. In the first type, the mean of the estimated values was 235·05, and the sum of the squared deviations from this was 0·3850; in the second type, the figures were 234·61 and 0·2084. Show that it is reasonable to conclude there was some systematic (as opposed to random) error in one or both experiments, assuming that the errors in each are normally distributed.

14 Fifteen dry cells of the same make were found to have mean voltage 1·495, with the sum of the squared deviations from this figure equal to 0·0237; for twenty cells of the same type but a different make, the figures were 1·545 and 0·0796. Assuming that the distributions of voltage are normal, what conclusions may be drawn?

15 A certain steel component is required to have a mean breaking load of at least 5600 newtons. From each of two batches of components 7 are selected at random, and the test results are as shown below.

Load (N) $\begin{cases} \text{Batch A} & 4515 \quad 5305 \quad 5665 \quad 4220 \quad 5045 \quad 5755 \quad 5020 \\ \text{Batch B} & 5100 \quad 5710 \quad 6080 \quad 5385 \quad 5425 \quad 5660 \quad 5315 \end{cases}$

(a) Assuming that the two batches are drawn from normal populations, discuss whether the population variances may reasonably be assumed to be the same, using the variance ratio test.

(b) Compare the average breaking loads of the two batches, using the t-test.

(c) It is pointed out to the manufacturer that the mean load overall is below the acceptable minimum, but he objects that the result is inconclusive because the sample is too small. What conclusion should an independent observer reach?

16 Tests were carried out on a system alleged to improve memory. Twenty individuals were divided into random groups of ten, and one group only taught the system. All were then given the same list of random digits to remember under identical conditions, and the number of digits remembered in their correct positions was recorded for each. The sum of the scores for the instructed group was 216, and the sum of the squared scores was 5498; for the control group the figures were 173 and 3554. What conclusions may be drawn?

Chapter 11

Correlation

The idea of correlation will be familiar to most readers. Does the boy who comes top of the class in French usually also do well in German? Is the number of O-level passes a reliable guide to a person's suitability for a certain job? Do tall men usually marry tall women? In each of these cases, our own experience would lead us to say that there is some measure of correlation between the two quantities. In other words, there is some degree of non-randomness when the two sets of figures are compared, assuming that it is possible to put the information into numerical form. In order to examine each case in more detail, we require a method of indicating the closeness of the correlation, which will be a measure of the amount of non-randomness. Before we attempt this, however, we must be aware of the limitations of our analysis.

Consider the following (fictitious) sets of figures, which represent the results of an experiment carried out to see if there is any connection between a person's weight and the amount of exercise he takes.

Individual:	A	B	C	D	E	F	G	H
Weight (kg):	55	65	67	73	75	82	89	92
Km walked per week, average:	62	51	53	48	44	41	37	30

These figures can be plotted on a graph of a type usually known as a **scatter diagram**. The result is shown in figure 11.1, and the points can be seen to lie roughly on a straight line.

There is evidently a considerable degree of correlation, and, from this fact, one of five possible conclusions can be drawn.
1 The sample was not a fair one.
2 The correlation could reasonably have been due to chance, and no conclusions can be drawn.
3 One can keep one's weight down by taking plenty of exercise.

E

4 Those who take more exercise than others generally do so because they are more lightly built. (Compared with no. 3, cause and effect are reversed.)
5 The correlation is due to some other untested factor. For example, well-off people might eat more heavily and travel more by car than those less well-off.

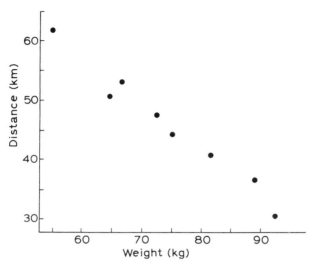

Figure 11.1   Scatter diagram showing weights and distances

The statistician's job is first to see that the sampling is free of bias, and then to decide whether the degree of correlation is enough to exclude, within reasonable limits, the possibility that it might have been due to chance. Considerations of cause and effect are usually outside his scope: even when it is apparently obvious which is which, he should resist the temptation to say so. It may well be that there is a high correlation between smoking and lung cancer, but to the statistician this might as easily be used to prove that incipient lung cancer makes people start smoking, as that smoking causes lung cancer.

When two variables are compared over a period of time, quite high correlation can often be obtained in the most unlikely circumstances. (G. U. Yule pointed out that there was close correlation between the suicide rate and membership of the Church of England.) This is simply because variations of both are governed by the same economic, social or other conditions. Another example, in which the underlying cause is more obvious, is the price of corn and the incidence of hay fever.

We shall now develop a quantitative measure of the degree of correlation in a set of paired observations. Such a set, by the way, is known as a **bivariate distribution**. Later, we shall be able to use this to test whether a given degree of correlation is significant.

In the following sets of figures, the first set is compared in turn with each of the other four. (For convenience, the figures in the reference set are evenly spaced, but the correlation does not depend on this.)

Set X:	$x =$	12	14	16	18	20	22	24	26	28
Set A:	$y =$	1	2	3	4	5	6	7	8	9
Set B:	$y =$	9	8	7	6	5	4	3	2	1
Set C:	$y =$	3	1	5	2	6	4	9	7	8
Set D:	$y =$	7	2	4	9	3	6	1	8	5

Take set A with X first. Here there is an exact linear relationship, which is in fact $y = \frac{1}{2}x - 5$. We call this perfect positive correlation—positive because increasing values of y are associated with increasing values of x. (Also, the straight line represented by the equation above has a positive gradient.) Similarly, set B has perfect negative correlation with X, the equation being $y = -\frac{1}{2}x + 15$. Set C has some positive correlation with X, as the larger numbers in C are generally towards the right-hand end— this could also be seen on a scatter diagram, in which the closeness of points to a line of best fit is a measure of the degree of correlation. For set D there is no obvious correlation.

The fact that the reference set consists of larger numbers than the others does not affect the degree of correlation; it would not matter, in fact, if we added or subtracted any number consistently in any one set. It actually makes the analysis easier if we subtract the mean of each set from every figure: the table now consists of deviations from the respective means.

Set X:	$d_x =$	-8	-6	-4	-2	0	2	4	6	8
Set A:	$d_y =$	-4	-3	-2	-1	0	1	2	3	4
Set B:	$d_y =$	4	3	2	1	0	-1	-2	-3	-4
Set C:	$d_y =$	-2	-4	0	-3	1	-1	4	2	3
Set D:	$d_y =$	2	-3	-1	4	-2	1	-4	3	0

You will notice that, for high positive correlation, the large positive quantities in one set are, generally speaking, paired off with the high positives in the other, and the same is true of the negative ones. Where there is high negative correlation, the high positives are paired off with the numerically high negatives, and similarly with the low figures; and where there is little correlation, there is no obvious pairing off. All this suggests a numerical measure for the degree of correlation: $\Sigma d_x d_y$.

The quantity $d_x d_y$ is tabulated below for the figures we have been using, and it will be seen that $\Sigma d_x d_y$ is large and positive for high positive correlation; large and negative for high negative correlation; and, for no correlation, consists of positive and negative quantities in random order. (On this last point, compare the proof of equation 8.1, on page 90.)

X with A: $d_x d_y =$	32,	18,	8,	2, 0,	2,	8,	18,	32;	$\Sigma d_x d_y =$	120
X with B: $d_x d_y =$	-32,	-18,	-8,	-2, 0,	-2,	-8,	-18,	-32;	$\Sigma d_x d_y =$	-120
X with C: $d_x d_y =$	16,	24,	0,	6, 0,	-2,	16,	12,	24;	$\Sigma d_x d_y =$	96
X with D: $d_x d_y =$	-16,	18,	4,	-8, 0,	2,	-16,	18,	0;	$\Sigma d_x d_y =$	2

A little thought will convince you that 120 is the maximum possible with these figures, so that, if we divide by 120, we have a measure of the correlation which is equal to ± 1 for perfect positive or negative correlation, and varies appropriately in between. For set C, for example, it equals 0·8, representing fairly high positive correlation. In order to have a universal formula, we only need to know how the constant of 120 is found, and at this point we go back to the beginning again. We shall define a coefficient known as the **product moment correlation coefficient** (or, more simply, as the **correlation coefficient**), which is, in fact, the quantity we have used above, and see what properties it has.

The product moment correlation coefficient

This is denoted by the letter r, and is defined by the equation

$$r = \frac{(1/n)\Sigma d_x d_y}{s_x s_y} \tag{11.1}$$

The numerator, which is the mean product of the deviations of each paired member of the distribution, is called the **covariance** of the bivariate distribution. The denominator $s_x s_y$ is the product of the SD's of the two parts of the distribution considered separately. The formula can also therefore be written

$$r = \frac{(1/n)\Sigma d_x d_y}{\sqrt{\{(1/n)\Sigma d_x^2 \times (1/n)\Sigma d_y^2\}}} = \frac{\Sigma d_x d_y}{\sqrt{(\Sigma d_x^2 \Sigma d_y^2)}} \tag{11.2}$$

(The word 'moment' in the name refers to the fact that deviations from a mean are sometimes known as moments; in mechanics, the distance of a unit force from a centre of rotation is a measure of its 'moment' or turning effect.)

Consider the bivariate distribution (x_1, y_1), $(x_2, y_2) \ldots (x_n, y_n)$. Let the mean be $(\bar{x}, \bar{y})$, and write $d_x = x - \bar{x}$, $d_y = y - \bar{y}$. We define perfect linear correlation by the condition that the equation $y = hx + k$ holds for every member of the distribution, where h and k are constants. (Compare the numerical example above.)

This gives, by addition, $\Sigma y = h\Sigma x + nk$, i.e. $\bar{y} = h\bar{x} + k$. Hence $y - \bar{y} = h(x - \bar{x})$, and so $d_y = hd_x$. So, in this case,

$$r = \frac{(1/n)\Sigma d_x d_y}{\sqrt{\{(1/n)\Sigma d_x^2 \times (1/n)\Sigma d_y^2\}}} = \frac{\Sigma(d_x \times hd_x)}{\sqrt{(\Sigma d_x^2 \times \Sigma h^2 d_x^2)}} = 1 \quad \text{if} \quad h > 0.$$

If $h < 0$, the numerator is negative, and the positive value of the square root is still used, so $r = -1$. The formula as defined thus gives the required values of ± 1 for perfect positive and negative correlation, and can be used between these values to measure the degree of correlation.

The two sets of figures may be multiplied by any positive constants, if it helps in the calculation process, without affecting the answer: thus any units can be used, and numbers with decimal or fractional parts may be transformed into whole numbers. The proof of this property is not difficult. Suppose that the values of d_x and d_y in the formula are multiplied by u and v respectively. Then

$$r = \frac{(1/n)\Sigma ud_x vd_y}{\sqrt{\{(1/n)\Sigma u^2 d_x^2 \times (1/n)\Sigma v^2 d_y^2\}}} = \frac{(1/n)\Sigma d_x d_y}{\sqrt{\{(1/n)\Sigma d_x^2 \times (1/n)\Sigma d_y^2\}}} \quad \text{as before.}$$

If $\bar{x}$ and $\bar{y}$ are exact round figures, the calculation of r presents no difficulty, and only an additional column for $d_x d_y$ is required compared with the usual layout for the calculation of a SD. Where such fortunate coincidences do not happen, a formula using working means is required. Since the denominator is the product of two SD's, no new ideas are involved here, and it is only necessary to find an alternative form for $\Sigma d_x d_y$.

The working means are denoted by x_0 and y_0, and let $\bar{x} - x_0 = \delta_x$ and $\bar{y} - y_0 = \delta_y$. The quantities d_x and d_y are now deviations from the working means. So, from equation 2.3, $\bar{x} = x_0 + \Sigma d_x/n$ and $\bar{y} = y_0 + \Sigma d_y/n$. The term previously written as $\Sigma d_x d_y$ now becomes

$$\Sigma(x - \bar{x})(y - \bar{y}) = \Sigma(x_0 + d_x - \bar{x})(y_0 + d_y - \bar{y})$$

$$= \Sigma(d_x - \delta_x)(d_y - \delta_y)$$

$$= \Sigma d_x d_y - \delta_x \Sigma d_y - \delta_y \Sigma d_x + \Sigma \delta_x \delta_y$$

$$= \Sigma d_x d_y - \delta_x \times n\delta_y - \delta_y \times n\delta_x + n\delta_x \delta_y$$

$$= \Sigma d_x d_y - (1/n)\Sigma d_x \Sigma d_y.$$

The formula for the calculation of r, using working means, is therefore

$$r = \frac{(1/n)\Sigma d_x d_y - (1/n^2)\Sigma d_x \Sigma d_y}{\sqrt{\{[(1/n)\Sigma d_x^2 - (\Sigma d_x/n)^2][(1/n)\Sigma d_y^2 - (\Sigma d_y/n)^2]\}}} \qquad (11.3)$$

It will help you to remember the numerator if you notice that putting $d_x = d_y$ makes it into the usual form for a variance. Because of this, the form above is the easiest one to remember, if you are unfortunate enough to have to remember it. But actual calculations are easier if the n's are rearranged thus:

$$r = \frac{\Sigma d_x d_y - (1/n)\Sigma d_x \Sigma d_y}{\sqrt{\{[\Sigma d_x^2 - (1/n)(\Sigma d_x)^2][\Sigma d_y^2 - (1/n)(\Sigma d_y)^2]\}}} \qquad (11.4)$$

Example 1 Find the correlation coefficient for the bivariate distribution listed under x and y below.

x	y	d_x	d_y	d_x^2	d_y^2	$d_x d_y$	$d_x - d_y$	$(d_x - d_y)^2$
21	5·6	−9	−14	81	196	126	5	25
25	6·7	−5	−3	25	9	15	−2	4
27	6·3	−3	−7	9	49	21	4	16
27	6·6	−3	−4	9	16	12	1	1
28	7·1	−2	1	4	1	−2	−3	9
31	6·8	1	−2	1	4	−2	3	9
33	7·6	3	6	9	36	18	−3	9
34	7·6	4	6	16	36	24	−2	4
34	9·1	4	21	16	441	84	−17	289
37	8·7	7	17	49	289	119	−10	100
$x_0 = 30$	$y_0 = 7$	−3	21	219	1077	415	−24	466

Notice that, in the d_y column, the decimal point has been omitted. The last two columns are for checks, using the identities $\Sigma(d_x - d_y) = \Sigma d_x - \Sigma d_y$ and $\Sigma(d_x - d_y)^2 = \Sigma d_x^2 - 2\Sigma d_x d_y + \Sigma d_y^2$. Here $-3 - 21 = -24$, and $219 - 2 \times 415 + 1077 = 466$. This checks everything except the d_x and d_y terms themselves, and so a separate check of each term by addition is advisable. After the checks are made,

$$r = \frac{10 \times 415 - (-3) \times 21}{\sqrt{\{(10 \times 219 - 9)(10 \times 1077 - 441)\}}} = \frac{4213}{\sqrt{(2181 \times 10\,329)}} = 0\cdot89.$$

A check using $(d_x + d_y)$ can also be used but, with positive correlation, it gives larger figures in the two columns. This version should be used when negative correlation is expected.

The significance of a correlation coefficient

There are three situations in which we may wish to test the significance of a correlation coefficient:
1 Where a value of r is to be compared with a value predicted theoretically, or a suggested population value.
2 When it is desired to find if a given value of r is significantly different from zero; this is a particular case of the one above.
3 Where values of r obtained from two separate experiments are to be compared.
Only the first two of these will be considered here, and without any mathematical justification. The reader who wishes to investigate further should consult Kendall and Stuart's *Advanced Theory of Statistics*, which is the source of the limiting value of n quoted below.

The distribution of the coefficient itself tends to the normal too slowly to be of much value. Fortunately, a simple transformation due to R. A. Fisher can be used to get round the difficulty. If the null hypothesis is that the population value of the coefficient is ρ (Greek rho), then the quantities z and ζ (Greek zeta) are defined by

$$z = \tfrac{1}{2}\log_e \frac{1+r}{1-r} \quad \text{and} \quad \zeta = \tfrac{1}{2}\log_e \frac{1+\rho}{1-\rho}.$$

The difference $z - \zeta$ is approximately normally distributed with zero mean and SE $\sigma_z = 1/\sqrt{(n-3)}$. This can be used with confidence for values of n down to 50 and, except in marginal cases, lower still. The transformation is tabulated in Table 8.

Example 2 Find the significance of $r = 0.4$ when $n = 50$. Here the null hypothesis is that $\rho = 0$. $z = \tfrac{1}{2}\log_e(1.4/0.6) = 0.4236$, and $\sigma_z = 1/\sqrt{47} = 0.1458$. The value of z is thus about $2.91\sigma_z$, giving a significance level (two tails) of about 0.4%.

Example 3 If three dice are thrown together, and the total score on A and B is compared with the total on A and C, the expected value of the correlation coefficient is $\tfrac{1}{2}$. In 100 such throws, a coefficient of 0.37 was obtained. Is this in agreement with the value given?

Here $\zeta = \tfrac{1}{2}\log_e 3 = 0.5493$ and $z = \tfrac{1}{2}\log_e (1.37/0.63) = 0.3883$. The difference of 0.161 has to be compared with σ_z, which is $1/\sqrt{97} = 0.1015$, and the result is not significant; that is, it does not cast doubt on the null hypothesis (*or* on the experiment).

Rank correlation

In certain types of test, objective measurements are impossible and the results are given in the form of an order of merit or preference. Examples are the kind of competitions often found on breakfast cereal packets. Where eight pictures are given an order of merit by two people independently, the results might appear as:

Picture labelled:	A	B	C	D	E	F	G	H
Position given by X:	3	6	5	8	4	1	7	2
Position given by Y:	5	8	3	7	6	1	2	4

A coefficient of correlation can be calculated from these figures and, since the positions in order are known as **ranks**, it is called a coefficient of rank correlation. The two usual methods were developed by C. Spearman and M. G. Kendall, and the coefficients R and τ (Greek tau) are named after them. Both vary between ± 1 like the product moment coefficient but, except at these limits, the values are not the same.

Rank coefficients may be used in correlating any two sets of figures if they are first put into rank order: for example, correlating two form orders for French and English rather than the actual marks from which the positions were derived. The effect of this procedure will be mentioned later.

Spearman's coefficient of rank correlation

This is simply the product moment coefficient using the ranks as the data. Since the figures in each set are the numbers from 1 to n in some order, a simplified form of calculation can be used.

In deriving the formula for Spearman's coefficient, two standard algebraic results are required, proofs of which will be found in most advanced algebra books. These are the sum of the first n integers, and the sum of their squares, which are

$$\sum_{x=1}^{n} x = \tfrac{1}{2}n(n+1) \quad \text{and} \quad \sum_{x=1}^{n} x^2 = \tfrac{1}{6}n(n+1)(2n+1).$$

The variance of the set of integers from 1 to n is therefore as follows, using equation 3.5:

$$\sigma^2 = (1/n)\Sigma x^2 - (\Sigma x/n)^2$$
$$= \tfrac{1}{6}(n+1)(2n+1) - \tfrac{1}{4}(n+1)^2$$
$$= \tfrac{1}{12}(n+1)(n-1) = \tfrac{1}{12}(n^2-1).$$

Since the denominator of the correlation coefficient consists of the product of two SD's, each of which is the square root of this variance, the denominator is $\frac{1}{12}(n^2-1)$.

The next step is to find the numerator, $(1/n)\Sigma d_x d_y$. Equation 11.3 is used for this, with a working mean of zero, giving

$$(1/n)\Sigma d_x d_y = (1/n)\Sigma xy - (1/n^2)\Sigma x\Sigma y.$$

We shall now express this in terms of the differences between ranks, $x-y$, which we shall call D. The expression can be written

$$(1/2n)\{\Sigma x^2 + \Sigma y^2 - \Sigma(x-y)^2\} - (1/n^2)\Sigma x\Sigma y.$$

Since $\Sigma x = \Sigma y$ and $\Sigma x^2 = \Sigma y^2$, this equals

$$(1/n)\Sigma x^2 - (\Sigma x/n)^2 - (1/2n)\Sigma D^2.$$

The first two terms of this are identical with the variance worked out above, and the whole expression simplifies to give

$$R = 1 - \frac{6\Sigma D^2}{n(n^2-1)}. \tag{11.5}$$

Example 4 Find Spearman's coefficient for the two rank orders below.

Order X:	3	6	5	8	4	1	7	2
Order Y:	5	8	3	7	6	1	2	4

Differences D:	-2	-2	2	1	-2	0	5	-2
D^2:	4	4	4	1	4	0	25	4; $\Sigma D^2 = 46$

$$\therefore\ R = 1 - \frac{6 \times 46}{8 \times 63} = 0\cdot45.$$

The simplicity of the calculation compared with that of the product moment coefficient does not need stressing. One small complication can arise: two or more measurements may be given equal ranks. The procedure is to give each position the mean of the tied ranks—thus two ranked equal fifth are given rank $5\frac{1}{2}$, and three ranked equal fifth are given rank 6. There is a possible objection to this procedure on the grounds that $5\frac{1}{2}^2$ is not the mean of 5^2 and 6^2, but the error is not large enough to bother about except in extreme cases, such as when there are only two or three distinguishable ranks. This is illustrated in exercise 8 at the end of this chapter.

The coefficient should normally be given to two decimal places, though the value of the second is often doubtful. For a significance test, in the case where the null hypothesis is that there is no correlation, R is approximately normally distributed with zero mean and variance $1/(n-1)$. This holds for values of n over about 20, and is only slightly affected by a few tied ranks.

Kendall's coefficient of rank correlation

This may be used as an alternative to Spearman's. It is a little more awkward to calculate, but it can be adapted to cope with data added after the initial calculation, and there are marginal advantages in testing its significance. Only a brief treatment is given here; for more details see Kendall's own book, *Rank Correlation Methods*, which gives a simple account of both coefficients as well as a full mathematical treatment. The calculation is easiest if one set of ranks is in counting order, and an example will be used to show the method.

Item labelled:	E	A	F	D	B	C
Rank given by X:	1	2	3	4	5	6
Rank given by Y:	3	6	5	1	4	2

The object is to count up the number of pairs, P, for which X and Y place the items in the same order—that is, counting EA but not AF. If the ranking orders are the same, P will take its maximum value of $\frac{1}{2}n(n-1)$, which is 15 here; and if the orders are exactly reversed, P will be zero.

The scoring pairs are EA, EF, EB, DB and DC, giving $P = 5$. Rather than by writing all the pairs down, the total can be counted up as follows: for each number in the Y-line in turn, count how many larger numbers appear on its right. You should follow this through to check that it is equivalent to what we want, and then try it out using the data of example 4, for which $P = 19$.

The coefficient is now calculated as follows. It is based on the value of P, and the formula is chosen in such a way that the coefficient has a range ± 1, as with Spearman's and the product moment coefficient. Thus

$$\tau = \frac{2P}{\frac{1}{2}n(n-1)} - 1 \qquad (11.6)$$

Here $n = 6$ and $P = 5$, so that $\tau = -\frac{1}{3}$. Spearman's R is -0.43 for comparison. For example 4, $\tau = 0.36$.

Tied ranks are dealt with as follows; as with Spearman, they are first averaged.

1	$2\frac{1}{2}$	$2\frac{1}{2}$	4	6	6	6	8	9	10
4	3	1	6	2	8	10	8	8	5

Where there is a tie in the lower line, $\frac{1}{2}$ instead of 1 is added to the score for P: thus the first 8 scores $1+\frac{1}{2}+\frac{1}{2} = 2$, and the second 8 scores $\frac{1}{2}$. Where there is a tie in the top line, count up P for the order as shown, then count again for the pair (or group) reversed, and take the mean score. Thus with the order printed above, the 3 scores 6 and the 1 scores 7; if the 3 and 1 are interchanged, the 1 scores 8 and the 3 still scores 6. The total contribution to P from these two is therefore $13\frac{1}{2}$. For the complete calculation, $P = 29\frac{1}{2}$ and $\tau = (59/45)-1 = 0\cdot31$.

If the preliminary ordering of the pairs is considered tedious, a direct method is possible. The figures used earlier, now in the order A to F, will serve to illustrate.

	1̸	2	3	4̸	5	6̸
X:	2	5	6	4	1	3
Y:	6	4	2	1	3	5

Above the data, the counting numbers are written as shown—the deletions are part of the method. The ranks given by Y are now taken in order. 1 in Y is paired with 4 in X; delete the 4 in the top line, and score 2 for the number of figures to the right of 4 in the top line. 2 in Y goes with 6 in X; delete 6, score nil. 3 in Y goes with 1 in X; delete 1, score 3, the deleted figures not being counted. Pause at this point to see why the method works.

We require the number of pairs (x, y) for which both $x > 1$ and $y > 3$—this is the basis of the previous method. This is equivalent to saying that we want, for the rank 3 in Y, the number of ranks in X which are greater than 1, excluding those which are paired with ranks less than 3 in Y. Now ranks less than 3 in Y have already been dealt with, and the corresponding ranks in X have been deleted in the top line. The number of undeleted figures in the top line to the right of a given rank in X is thus the contribution towards P for that pair of ranks.

In this example, there are no further scores, and $P = 5$ as before.

To find the significance of τ, when the null hypothesis is that there is no correlation, its numerical value must first be reduced by $\dfrac{1}{\frac{1}{2}n(n-1)}$ as a continuity correction. It is then normally distributed for $n \geqslant 10$, with zero mean and variance $\dfrac{4n+10}{9n(n-1)}$. Tied ranks matter little if there are only a few of them.

The two types of correlation coefficient compared

Where the ranks are not derived from objective measurements, rank coefficients are the only type which can be used. Where actual measurements are available, the product moment coefficient is perhaps better on the grounds that information is lost in the ranking procedure. But, since it is awkward to calculate (and since high accuracy is often unnecessary), a ranking method may well be an acceptable substitute. If the correlation is not linear, ranking methods may even give a better idea of the true degree of correlation. But if there appears to be close non-linear correlation, and if accuracy is important, a transformation of one or both sets of figures (e.g. logarithmic) may give the best results.

Exercises

1 Find the product moment correlation coefficient for the bivariate distribution below. (In this question and in numbers 2 and 3 the means are round numbers, so that equation 11.2 may be used.)

x: 23   17   14   27   21   18
y: 39   35   41   49   42   34

2 Repeat the instructions of question 1 for the distribution:

x: 19   32   27   24   29   21   33   23
y: 56   38   47   41   41   49   37   51

3 Repeat the instructions of question 1 for the distribution:

x: 5·1   8·4   5·6   4·5   6·7   4·6   6·1   7·8
y: 97   79   75   81   72   85   81   78

4 Find the product moment coefficient for the pairs of marks below. Then write down the two orders of merit, and find Spearman's and Kendall's coefficients for comparison.

x: 57   41   51   89   47   40   82   51   37   54   79   62
y: 59   50   62   76   39   42   79   68   48   70   91   54

5 Repeat the instructions of question 4 for the paired marks:

x: 37   47   83   75   62   53   59   67   61   66
y: 65   41   57   69   39   82   46   53   71   57

6 Find the product moment coefficient for the bivariate distribution:

x: 69   70   73   72   69   69   72   66   73   69   71   74   69   75   67   72
y: 58   54   54   80   50   64   75   20   50   54   49   52   35   54   12   54

7 Thirty car owners were asked to state (a) the year of registration of their cars and (b) the amounts they spent on mechanical repairs in the last twelve months, to the nearest pound. Cars less than a year

out of guarantee were excluded from the sample. The figures were as follows:

| Year, 19– | 68 | 69 | 70 | 69 | 67 | 63 | 70 | 70 | 69 | 70 | 65 | 68 | 67 | 68 | 70 |
| Repairs, £ | 12 | 6 | 17 | 0 | 5 | 13 | 6 | 2 | 7 | 3 | 12 | 20 | 11 | 7 | 9 |

| Year, 19– | 64 | 66 | 69 | 69 | 68 | 70 | 69 | 69 | 68 | 67 | 70 | 66 | 70 | 70 | 69 |
| Repairs, £ | 35 | 10 | 7 | 0 | 8 | 15 | 9 | 19 | 2 | 13 | 10 | 10 | 8 | 3 | 2 |

Find the correlation coefficient. Stating carefully all the assumptions you make, carry out a significance test, and comment on it briefly.

8 Calculate the product moment coefficient and Spearman's coefficient for the two rank orders below, in order to demonstrate that the approximation involved in using the means of tied ranks in Spearman's method is small—at least when there are not too many ties.

| 1 | $2\frac{1}{2}$ | $2\frac{1}{2}$ | 4 | 5 | 7 | 7 | 7 | 9 | 10 |
| 2 | 6 | 1 | 5 | 3 | 4 | 10 | 8 | 7 | 9 |

9 From tables of random digits, write down, for a number of groups of four digits, the sum of the first, second and third and the sum of the second, third and fourth, and determine the correlation coefficient. (This is best done by a group of people, in which case each person should use a working mean of 13, and the column totals should be added together for a single final calculation.) Test whether the coefficient so obtained is consistent with hypotheses that the mean of the probability distribution for the coefficient is 0, $\frac{1}{2}$, $\frac{2}{3}$ and $\frac{3}{4}$ in turn.

10 Prove that Spearman's coefficient for ranks 1 to n compared with n down to 1 is -1. (The proof is slightly different for the two cases n even and n odd; try n odd first.)

11 Draw graphs showing the minimum values of r which are significant at the 5 % and the 1 % levels, from $n = 20$ to $n = 1000$. (Use Table 8; below $n = 50$ the results will not be very accurate.)

12 Discuss in general terms how much correlation might be expected between paired measurements of the following:
(a) Heights and weights for the 20 members of a sixth form of a boys' school.
(b) Heights and weights for a random sample of 20 from a boys' school in which the age range is 11 to 18.
(c) Rank orders estimated by two people of the masses of ten differently shaped pieces of wood which are, in fact, of masses ranging in equal steps from 15·0 to 15·9 kg.
(d) Production of sports cars and imports of caviar, for the United Kingdom in the years 1936 to 1956.
(e) The sums and differences of pairs of random digits.

13 Determine correlation coefficients, and their significance where possible, from your own data. Here are some suggestions.

(a) Weights, ages up to about 16, and heights and other measurements, in any combination.

(b) Heights, or ages at marriage, of married couples.

(c) Comparisons of subject orders within a form, either with each other or with an overall order.

(d) Biological field measurements. There is plenty of scope here, e.g.: length and breadth of leaves; length of leaves with number of spines; girth and number of rings on cut timber; numbers of stamens and carpels on different types of flower.

14 Explain, as if to an intelligent person who knows nothing of analytical statistics, the term 'correlation' and the use of the two types of coefficient, in about 200 words.

Chapter 12

Estimation from related data

For certain types of steel there is quite high correlation between the ultimate tensile strength and the hardness. It is usually more important to know the strength than it is to know the hardness, because the result of low strength may be a fracture; but hardness is much easier to measure, as the test is non-destructive and requires less equipment. If we know how the correlation operates, it is possible to predict, from a given hardness measurement, the corresponding ultimate strength—as usual, within certain limits rather than as a precise figure. How it is done is the problem we shall consider in this chapter, first graphically and then analytically.

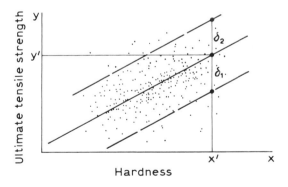

Figure 12.1   Scatter diagram for strength/hardness data

The essential first step is to draw a scatter diagram to represent the data. The notation and axes are chosen so that x is the variable we shall later use to estimate a value of y. (That is, x is the variable conventionally described as independent, though the word is not appropriate in statistical work.) This is important, because as we shall see the method is not directly

137

reversible. Then, if there is fairly high linear correlation, it will be possible to draw by eye a line of best fit, with the points scattered evenly on both sides of the line. Accuracy is not important at this stage, as we shall discuss the criteria for 'best' later. Then, for any value x of the hardness, the best estimate for the strength will be that marked y'.

If there is high correlation which is evidently not linear, a curve may be drawn and estimates based on it. But this is far more difficult to analyse mathematically, and we shall consider only the linear case.

The next stage is to form a rough idea of how accurate the estimate is. On the scatter diagram, two lines are drawn, parallel to the line of best fit and the same distance each side of it, in such a way as to include between them about 95 % of the observations. These are shown as broken lines above. Then, for a given x, we can say that the 95 % confidence limits for y are $y' \pm \delta$, where δ is measured as shown.

We can now begin to see how the mathematical analysis is going to take shape. First we shall find the equation of the line of best fit, which is called the **regression line of y on x**. This will give an unbiased estimate y' for a given x. Then we shall find the **standard error of estimate** σ_e of y', which is calculated from the deviations from the line in the y-direction— the quantity δ on the scatter diagram is roughly twice the SE. Finally, assuming that these deviations are normally distributed, we shall set up confidence limits for y' which will be either $y' \pm t\sigma_e$ or $y' \pm z\sigma_e$, depending on the size of the distribution.

Before starting the analysis, a few words of caution. It is essential that the correlation is very close to linear and, to ensure this, a scatter diagram must be used in conjunction with the calculations. Also, a regression line cannot be used reliably for extrapolation (that is, predicting results beyond the limits of the original data), at least without some theoretical justification.

The regression line of y on x

In chapter 3, we found (in the derivation of equation 3.4) that the sum of the squares of the deviations of a distribution from any point is a minimum when the chosen point is the mean. By analogy, we shall define the line of best fit here as the line which makes the sum of squares Σd^2 a minimum, where the d's are the deviations from the line measured in the y-direction, as shown in figure 12.2. (For the regression line of x on y, in which y is taken as the variable from which x is to be estimated, the d must be measured in the x-direction.) This assumption leads to two other results which might be predicted from symmetry: that the line passes through $(\bar{x}, \bar{y})$, and that Σd is zero.

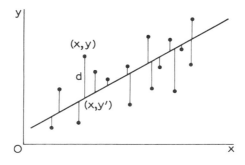

Figure 12.2   The regression line of y on x

Let the equation of the line be $y-\bar{y} = b(x-h)$. (This form is chosen because we shall later prove that $h = \bar{x}$.) Consider a member (x, y) of the distribution; this will not, in general, lie on the line. Let the estimate for y corresponding to x be y', so that (x, y') is the point where the ordinate through x meets the line. Then if $y-y' = d$, we have the condition that Σd^2 $(= S$, say) is a minimum.

Since (x, y') is on the line, $y'-\bar{y} = b(x-h)$, and so $d = y-y' = (y-\bar{y})-b(x-h)$. Hence

$$S = \Sigma d^2 = \Sigma\{(y-\bar{y})-b(x-h)\}^2$$
$$= \Sigma(y-\bar{y})^2 - 2b\Sigma(y-\bar{y})(x-h) + b^2\Sigma(x-h)^2$$

The minimum value of S has to be established by taking optimum values of the two quantities not so far fixed for a given distribution, h and b. First consider the effect of changing h, leaving b as a constant, which requires the condition $dS/dh = 0$. (This is partial differentiation, as some readers will know, but the notation for this has not been used as others will not be familiar with it.) Expanding the expression above,

$$S = \Sigma(y-\bar{y})^2 - 2b\Sigma x(y-\bar{y}) + 2bh\Sigma(y-\bar{y}) + b^2\Sigma x^2 - 2hb^2\Sigma x + nh^2b^2$$

$$\therefore\ dS/dh = 2b\Sigma(y-\bar{y}) - 2b^2\Sigma x + 2nhb^2 = 0$$

$$\therefore\ \Sigma(y-\bar{y}) = b(\Sigma x - nh), \quad \text{since } b \neq 0.$$

Since $\Sigma(y-\bar{y}) = 0$, $h = \Sigma x/n = \bar{x}$. The line therefore passes through $(\bar{x}, \bar{y})$, which is known as the **mean centre** of the bivariate distribution. It also follows that $\Sigma d = \Sigma(y-\bar{y}) - b\Sigma(x-\bar{x}) = 0$, as mentioned earlier.

We can now rewrite the earlier equation for S, first putting $\bar{x}$ for h, and then writing d_x and d_y for the deviations from the mean in the separate x- and y-distributions. We can then find b from the condition $dS/db = 0$. Hence

$$S = \Sigma(y-\bar{y})^2 - 2b\Sigma(y-\bar{y})(x-\bar{x}) + b^2\Sigma(x-\bar{x})^2$$

$$= \Sigma d_y^2 - 2b\Sigma d_x d_y + b^2\Sigma d_x^2$$

$$\therefore dS/db = -2\Sigma d_x d_y + 2b\Sigma d_x^2 = 0$$

$$\therefore b = \frac{\Sigma d_x d_y}{\Sigma d_x^2} = \frac{(1/n)\Sigma d_x d_y}{s_x} \tag{12.1}$$

The full equation of the regression line of y on x is therefore

$$y - \bar{y} = \frac{\Sigma d_x d_y}{\Sigma d_x^2}(x - \bar{x}) \tag{12.2}$$

The quantity b, which is the gradient of the regression line, is known as the **regression coefficient** of y on x. The method of calculation follows that of the correlation coefficient. As a method of remembering that it is Σd_x^2 in the denominator rather than Σd_y^2, notice that, if you 'cancel' the Σ sign and d_x, you are left with d_y/d_x which is rather like the usual form for a gradient.

We now have an unbiased estimator for y on given x, and it remains to calculate the SE of estimate σ_e. By definition, $\sigma_e^2 = \Sigma d^2/(n-1) = S/(n-1)$, using divisor $n-1$ because we require an estimate for data beyond the sample, i.e. a population estimate. We have, therefore,

$$S = \Sigma d_y^2 - 2b\Sigma d_x d_y + b^2\Sigma d_x^2$$

$$= \Sigma d_y^2 - 2b^2\Sigma d_x^2 + b^2\Sigma d_x^2$$

$$= \Sigma d_y^2 - b^2\Sigma d_x^2 = \Sigma d_y^2\left\{1 - \frac{(\Sigma d_x d_y)^2}{\Sigma d_x^2 \Sigma d_y^2}\right\}$$

$$= \Sigma d_y^2(1 - r^2)$$

$$\therefore \sigma_e^2 = S/(n-1) = \frac{\Sigma d_y^2}{n-1}(1 - r^2) = \hat{\sigma}_y^2(1 - r^2)$$

$$\therefore \sigma_e = \hat{\sigma}_y\sqrt{(1 - r^2)}. \tag{12.3}$$

The quantity r is, of course, the correlation coefficient as defined by equation 11.2.

A confidence interval can be set up from this SE in the usual way, assuming that the distribution of the errors in y is normal. The formula for $n > 100$ is $y = \bar{y} + b(x - \bar{x}) \pm z\sigma_e$, where z is the appropriate normal variate for the confidence level required. For smaller values of n, the t-distribution is used, for $n - 2$ degrees of freedom.

If no information about x were available, the SE of the estimate for y would be the estimated SD itself, $\hat{\sigma}_y$; if there is some correlation, the SE is reduced by multiplying it by $\sqrt{(1 - r^2)}$, which grows smaller for increasing correlation, as might be expected. For example, if the correlation coefficient is $0 \cdot 8$, the confidence interval for y at a given significance level is reduced to 60% of its former size—a useful, but not spectacular, improvement.

A regression line should never be used only to state a point estimate without qualification. At the very least a scatter diagram should be given as well, so as to show clearly the kind of error to which the estimate is subject. Alternatively, the correlation coefficient may be given if a SE of estimate is not possible; if the correlation is not significant, a regression line should not be used at all.

Example 1 Eleven candidates took an examination in two similar parts, and a further candidate scored 48 in the first part but was absent from the second part. The 11 paired marks were those shown in columns x and y below. Use a regression line to give an unbiased estimate and a 95% confidence interval for the missing mark.

x	y	d_x	d_y	$d_x - d_y$	d_x^2	d_y^2	$d_x d_y$	$(d_x - d_y)^2$
36	47	-24	-13	-11	576	169	312	121
40	39	-20	-21	1	400	441	420	1
49	45	-11	-15	4	121	225	165	16
51	56	-9	-4	-5	81	16	36	25
58	53	-2	-7	5	4	49	14	25
60	61	0	1	-1	0	1	0	1
63	68	3	8	-5	9	64	24	25
68	64	8	4	4	64	16	32	16
68	74	8	14	-6	64	196	112	36
71	75	11	15	-4	121	225	165	16
78	69	18	9	9	324	81	162	81
$x_0 = 60$	$y_0 = 60$	-18	-9	$-9\checkmark$	1764	1483	1442	$363\checkmark$

Hence $\bar{x} = 58 \cdot 36$ and $\bar{y} = 59 \cdot 18$. Using equation 11.4,

$$r = \frac{1442 - \frac{1}{11}(-18)(-9)}{\sqrt{\{(1764 - \frac{324}{11})(1483 - \frac{81}{11})\}}} = \frac{1427 \cdot 3}{\sqrt{(1734 \cdot 5 \times 1475 \cdot 6)}}$$

$$= 0 \cdot 8920,$$

and, by the equivalent form of equation 12.1,

$$b = \frac{1427 \cdot 3}{1734 \cdot 5} = 0 \cdot 8230.$$

The equation of the regression line of y on x is, therefore, from equation 12.2,

$$y - 59 \cdot 18 = 0 \cdot 8230(x - 58 \cdot 36)$$

i.e. $$y = 0 \cdot 823x + 11 \cdot 1.$$

So the best estimate of y for $x = 48$ is 50·6, say 51 marks.

Notice that the last step involves subtraction between two large quantities, which is the reason why full four-figure accuracy is used up to that point.

Now the confidence interval. The unbiased estimate for the population variance of the y-distribution is

$$\frac{1}{n-1}\left\{\Sigma d_y^2 - (1/n)(\Sigma d_y)^2\right\} = \frac{1475 \cdot 6}{10} = 147 \cdot 56.$$

Hence

$$\sigma_e = \sigma_y\sqrt{(1-r^2)} = \sqrt{\{147 \cdot 56(1 - 0 \cdot 8920^2)\}} = 12 \cdot 15 \times 0 \cdot 204 = 2 \cdot 48.$$

For 95% confidence and 9 d.f., $t = 2 \cdot 262$, and $t\sigma_e = 5 \cdot 6$. Hence the 95% confidence interval for the missing mark is $50 \cdot 6 \pm 5 \cdot 6$, i.e. 45 to 56 marks.

The two regression lines

For the regression line of y on x, the coefficient b can be written in the form

$$b = \frac{\Sigma d_x d_y}{\Sigma d_x^2} = \frac{\Sigma d_x d_y}{\sqrt{(\Sigma d_x^2 \Sigma d_y^2)}}\sqrt{\left(\frac{\Sigma d_y^2}{\Sigma d_x^2}\right)} = r\frac{s_y}{s_x}.$$

For the other line, that of x on y, the coefficient b' is given by

$$b' = \frac{(1/n)\Sigma d_x d_y}{s_y^2}$$

by transposing x and y in equation 12.1, and this time

$$b' = r\frac{s_x}{s_y}.$$

This line also passes through $(\bar{x}, \bar{y})$, and its gradient (with the axes the usual way round) is $1/b'$.

Two results follow at once. First, $bb' = r^2$. Second, the two lines are identical if and only if $r^2 = 1$, i.e. the case of perfect correlation. For, since the lines intersect at $(\bar{x}, \bar{y})$, the condition that they are identical is that they have the same slope, that is, $b = 1/b'$ or $r^2 = 1$.

A further consequence is that, if x_1 gives a best estimate y_1, then the best estimate of x, given y_1, is not x_1, unless, of course, $x_1 = \bar{x}$. In the figures for example 1, 48 for x gave 50·5 for y, but if the other regression line is calculated, it will be found that 50·5 for y gives a best estimate of almost exactly 50 for x. The estimates, in fact, always *regress* or *revert* towards the mean, which is the reason for the name given to the lines.

The angle between the lines can be calculated as follows, though the result is of limited interest for our purposes. Since the slopes are b and $1/b'$, the angle is

$$\tan^{-1}\left|\frac{b-(1/b')}{1+(b/b')}\right| = \tan^{-1}\left|\frac{bb'-1}{b'+b}\right|$$

$$= \tan^{-1}\left|\frac{r^2-1}{r\left\{\dfrac{s_x}{s_y}+\dfrac{s_y}{s_x}\right\}}\right|. \qquad (12.4)$$

This shows that, for no correlation, the lines are perpendicular, though this can also be shown directly. Even when the correlation is quite high, the angle is not negligible—the figure for the data of example 1 is about $6\frac{1}{2}°$. For $r = \frac{1}{2}$ the figure is about 37° if $s_y = s_x$.

Exercises

1 Determine the equation of the regression line of y on x for the bivariate distribution below, and use it to give a best estimate of y when $x = 53$.

x:	35	40	45	50	55	60
y:	93	84	78	70	59	52

2 Determine the equation of the regression line of y on x for the bivariate distribution below, and use it to give a best estimate of y when $x = 28$.

x:	22	35	30	27	32	24	36	26
y:	44	62	53	59	59	51	63	49

3 From 20 pairs of values (x, y), the following quantities are calculated: $\Sigma x = 400$, $\Sigma y = 220$, $\Sigma x^2 = 8800$, $\Sigma y^2 = 2620$, $\Sigma xy = 4300$. Find

the linear regression lines of y on x and x on y. Which would be more useful if (a) x is the age in years and y is the reaction time in milliseconds of 20 people; (b) x is the cost in thousands of pounds and y is the floor space in hundreds of square metres of 20 buildings?   (OL)

4 The length y centimetres of a metal rod was measured at various temperatures $t°$ Centigrade, giving the following results:

t: 105·0   110·0   115·0   120·0   125·0   130·0
y: 20·01   20·08   20·18   20·20   20·43   20·51

By means of a linear regression equation, estimate the length of the rod when the temperature is 122·5° Centigrade.   (JMB)

5 The following table gives the height x hundred metres above sea level, to the nearest unit of x, of twelve places in Switzerland, and the early morning temperature $y°$ Centigrade on the same day in August. Find the equations of the regression lines and the correlation coefficient between x and y.

x: 11   15   10   5   4   5   9   12   3   4   18   16
y: 13   4   10   18   14   13   5   9   18   18   8   6

(AEB)

6 Thirteen candidates took an examination in two similar parts, and a fourteenth candidate scored 61 in the first part but was absent from the second part. The 13 paired marks were as shown below. Use a regression line to make a fair estimate for the missing mark, and find also the 95 % confidence interval, assuming that the marks follow a normal distribution.

First part: 32   34   41   48   55   59   59   68   70   74   78   83   84
Second part: 41   38   50   43   49   52   60   65   63   79   71   79   77

7 Find the equations of the two regression lines for the bivariate distribution below, and plot the results (together with the seven observations) on a graph. (The graph illustrates the fact that the lines do differ by an appreciable amount, even when the correlation is quite high— it is about –0·96 here. It may also be used to confirm the 'regression' effect discussed in the chapter.)

x: 13   17   24   27   36   37   46
y: 59   51   53   47   42   43   37

8 A red die and a green are thrown together, the score on the red being denoted by x and the total score by y. For each possible value of x, find the mean of the probability distribution for y, and hence find the regression line of y on x. Find, similarly, the regression line of x on y, and hence the expected value of the correlation coefficient between x and y.

9 If x is the sum of three random digits, and y is the sum of the second

and third of these together with a fourth random digit, find the regression line of y on x, and (by symmetry) the regression line of x on y. Hence find the expected value of the correlation coefficient between x and y. (Compare exercise 9 of chapter 11.)

10 From 50 pairs of values (x, y), the following quantities were calculated: $\Sigma x = 1150$, $\Sigma y = 2850$, $\Sigma x^2 = 30\,500$, $\Sigma y^2 = 175\,250$, and $\Sigma xy = 71\,310$. Find the regression line of y on x, and the standard error of estimate of y. Hence, give the 95 % confidence limits for y when $x = 32$, assuming that the errors of y are normally distributed.

Chapter 13

The chi-square distribution

In chapter 9 we investigated the significance test for a proportion; in this chapter we shall look at a test for several proportions at once. The test for a proportion was essentially a comparison between a probability distribution and an observed distribution where the number of possible values of the random variable was two—this test does the same for any number of values, and so is far more general. It has been relegated to near the end of the book because the mathematical analysis of the situation is far too difficult to tackle at this level, and so only a brief outline of what to do in using the test can be given. Some examples will make the method clear.

Example 1 Three people were each given an ordinary die and asked to record the results of 600 throws. Comment on their figures, which were:

Score:		1	2	3	4	5	6
Frequency recorded by:	A:	98	100	102	99	99	102
	B:	109	98	92	103	108	90
	C:	93	86	121	91	127	82

A's results might be thought reasonable enough, but there is just a suspicion that they are too good to be true—perhaps he thought the experiment a waste of time, and just wrote down some figures near 100 for each? B shows greater variation—is the amount typical for an unbiased die, or is there a possibility of bias? C's die appears to show a bias towards 3 and 5—is this significant or not? These are the questions we shall attempt to answer.

For each possible category, we have an actual number of observations (f) and a predicted value from the probability distribution (ϕ), which is here 100 each time. (There seems to be no generally accepted notation for these, but f seems the obvious choice for the observed frequency,

146

and ϕ (Greek phi) is the nearest Greek equivalent. Alternatively, O for observed and E for expected frequency may be used.) The procedure is then to calculate

$$\Sigma \frac{(f-\phi)^2}{\phi}$$

The distribution of this quantity is known as the χ^2 (chi-square, pronounced ky-square) distribution; χ^2 is a quantity which has a probability distribution, and we are going to test the significance of a particular value of it, which will be denoted by X^2. The distribution (like that of t) depends on the number of degrees of freedom, which is here 5, since 5 of the 6 frequencies determine the last one for a given total. The working is set out typically as follows.

B's frequency, f:	109	98	92	103	108	90
Expected frequency, ϕ:	100	100	100	100	100	100
$\dfrac{(f-\phi)^2}{\phi}$	0·81	0·04	0·64	0·09	0·64	1·00

Hence $X^2 = \Sigma \dfrac{(f-\phi)^2}{\phi} = 3\cdot22$. Comparative figures are 0·14 for A and 18·20 for C. In order to find the significance levels, we look at Table 10 under 5 d.f. The figures given are for a one-tail test, and represent the probability that the particular value of χ^2 will be exceeded. Thus there is only a very small probability—much less than 1%—that a value as low as 0·14 would be obtained in a fair experiment, and so A almost certainly cooked his figures. At the other end of the scale, C's figure of 18·2 is very significant, at a level of less than 1%, and so his die was almost certainly biased. B's results are quite typical for an unbiased die.

Example 2 730 deaths from road accidents in a year occurred as in the table below, which shows the number of days on which each number of deaths was recorded. Do the figures follow a Poisson distribution within reasonable limits?

Number of deaths:	0	1	2	3	4	5	6	7	8	9
Number of days, f:	56	102	94	56	29	19	5	1	2	1

In order to carry out the test, the small figures at the tail must be grouped together. The way in which X^2 is calculated means that, when ϕ is small, a fairly small difference $f-\phi$ can make a large contribution to X^2. To some extent this is taken into account in the theory of the distribution, but errors start to creep in for values of ϕ less than about

10. It is generally assumed that one or two values as low as 5 are permissible, but that, otherwise, figures under 10 should be grouped. The value of X^2 is, of course, affected, but so is the number of d.f., and one of the useful features of this method is that the theory allows for this grouping process. But one cannot take two groups which look suspiciously low and combine them together.

In the Poisson expected frequency distribution for $\mu = 2$, the number of days falls below 10 for six deaths per day, and so the days with six or more are combined together. The two distributions are then as follows.

f:	56	102	94	56	29	19	9
ϕ:	49·4	98·8	98·8	65·8	32·9	13·2	6·1
$\dfrac{(f-\phi)^2}{\phi}$:	0·89	0·10	0·23	1·46	0·46	2·55	1·38

The value of X^2 is therefore 7·07.

The number of d.f. is 5 not 6, because the probability distribution has two distinct parameters which are arranged to be the same as those of the given distribution—the total number of observations, and the mean. If you are doubtful of this, think about it in another way.

If there were only two categories (instead of 7) with, say, f_1 and f_2 observations, and the expected distribution was ϕ_1 and ϕ_2, then $\phi_1 + \phi_2$ must equal $f_1 + f_2$, the total number of observations, and also $f_1 x_1 + f_2 x_2 = \phi_1 x_1 + \phi_2 x_2$, because the means have to be the same. These two equations are sufficient to determine ϕ_1 and ϕ_2, and in fact $f_1 = \phi_1$ and $f_2 = \phi_2$. If the two distributions are identical, there can be no significance test, or in other words, there are no degrees of freedom. When the number of categories goes up to 3, there is one degree of freedom, and so on.

(When this method is used for a normal distribution, the number of d.f. is *three* less the number of categories because, in addition to the two restrictions mentioned above, the SD of the probability distribution has to agree with the observed distribution.)

To return to the example, the value of X^2, 7·07, is less than the 5% value for 5 d.f., and so the Poisson distribution does give a fair representation of the data. As usual, this does not prove that the Poisson can always be used for similar data: there is, for instance, an indication in the original figures that the frequencies for the top end of the distribution are on the high side, and if this were due to the effect of multiple deaths in a single accident, the Poisson would certainly not apply.

Since we have just mentioned the restriction on the minimum frequency in a single category, one other restriction on the use of the χ^2 distribution may be mentioned here. The theory of the distribution makes use of the normal approximation to the binomial, and so it must not be used if the total number of observations is too small—a minimum of 50 is generally recommended.

Example 3 A coin shows 124 heads in 200 throws. Is it biased?
This is the simplest case, with 1 d.f., in which either the χ^2 distribution or the simple significance test for a proportion can be used, and we shall use both methods in turn.

There is one difference in using χ^2 in the special case of 1 d.f.: a correction for continuity, known as Yates' Correction, must be applied. As with a significance test using the normal approximation to the binomial, the effect is to decrease by $\frac{1}{2}$ the numerical values of the differences. In this case X^2 is $23\frac{1}{2}^2/100 + 23\frac{1}{2}^2/100$ or 11·04, which is significant at the 0·1 % level. The coin is almost certainly biased.

The SE of the proportion is $\sqrt{(\frac{1}{2} \times \frac{1}{2}/200)} = 0.0353$, and so the actual difference of 0·1175 (allowing for a continuity correction again) is equal to 3·33 SE's. On a two-tail test, this is likewise significant at the 0·1 % level.

Contingency tables

A common experimental procedure is to investigate the differences between observed frequencies in two or more groups—typically, the proportion of cures using different drugs on two groups of patients. The data is set out in a form known as a contingency table, and the significance of the differences between the proportions can be tested using the χ^2 distribution.

Example 4 Two drugs used on comparable groups of patients gave results as in the table below. Is type A better than type B?

	Cures	Deaths	Totals
Treatment A:	112	25	137
Treatment B:	80	34	114
Totals:	192	59	251

The method is to set up a similar table with the same row and column totals, on the assumption that there is no difference between the two treatments. Thus if 59 out of 251 deaths occur overall, a fraction $\frac{59}{251}$

might have been expected to die in each group under these conditions. This gives:

	Cures	Deaths	Totals
Treatment A:	104·8	32·2	137
Treatment B:	87·2	26·8	114
Totals:	192	59	251

Notice that, as soon as one of the four figures has been found, the others are determined by the known values for the row and column totals. A 2×2 contingency table thus has 1 d.f. We now calculate X^2 as before, using Yates' Correction.

$$X^2 = \frac{6 \cdot 7^2}{104 \cdot 8} + \frac{6 \cdot 7^2}{32 \cdot 2} + \frac{6 \cdot 7^2}{26 \cdot 8} + \frac{6 \cdot 7^2}{87 \cdot 2} = 4 \cdot 01.$$

This is significant at the 5 % level, and so it appears that A is more successful than B.

Where only two proportions are involved, the χ^2 test is an alternative to another significance test—this time the test of the difference between two proportions, described in chapter 10.

One of the difficulties of statistical work in medicine may be mentioned here. For a test such as this one to be valid, it is essential that the two groups are truly comparable—ideally, of course, they should be chosen randomly. But the necessary choice, whether deliberate or random, is often incompatible with a doctor's ethical standards. If, in this case, A is a new drug and B and established one, and if, on past experience, the most serious cases would be likely to die if given drug B, is it fair to withhold the possibly better type A in the interests of a statistical test? Needless to say, a doctor's first duty is to his patients, and the statistician must take second place.

Example 5 Five hundred adults were asked to state their preference in types of holiday—touring, at a coastal resort or at an inland resort— and also what political party they supported. Is there evidence of association between the preferences?

Choice of holiday:	Tour	Coast	Inland	Totals
Party: Progressives:	52*	79*	32	163
Democrats:	84*	97*	60	241
Nationalists:	41	37	18	96
Totals:	177	213	110	500

The four figures marked with an asterisk, together with the row and column totals, are sufficient to determine the other figures, and so there are 4 d.f. (In general, for an $h \times k$ contingency table, there are $(h-1)(k-1)$ degrees of freedom.) As before, we set up a table on the assumption that there is no difference between the proportions. The values of $(f-\phi)$ are shown in brackets.

	Tour	Coast	Inland	Totals
Progressives:	57·7 ($-5·7$)	69·4 (9·6)	35·9 ($-3·9$)	163
Democrats:	85·3 ($-1·3$)	102·7 ($-5·7$)	53·0 (7·0)	241
Nationalists:	34·0 (7·0)	40·9 ($-3·9$)	21·1 ($-3·1$)	96
Totals:	177	213	110	500

As an example of the calculation, the proportion of Progressives overall is 163/500, and this, therefore, is the expected proportion of the 177 who prefer tours who are also Progressives. As a check on the working, the sum of each row and column of $(f-\phi)$ is zero. The value of X^2 from this table is 5·85, which is not significant. It may be true that Progressives prefer the seaside, but the evidence of the data is not sufficient to justify such a conclusion.

Exercises

1 Use the χ^2-distribution to test the significance of a single proportion, with the data of exercises 7 to 13 of chapter 9.

2 Set up contingency tables to test the significance of the difference between two proportions using the χ^2-distribution, with the data of exercises 1 to 6 of chapter 10.

3 At a public library during a given week, the following numbers of books were borrowed. Is there reason to believe that more books are generally borrowed on one weekday than another?
Day: Mon. Tues. Wed. Thurs. Fri. Sat.
Number issued: 204   292   242   283   252   275   (AEB)

4 Four coins were tossed 4096 times, and the number of heads was noted each time, with the results shown below. Examine the hypothesis that the coins were unbiased.
Number of heads: 0   1   2   3   4
Observed frequency: 249 1000   1552   1050   245   (AEB)

5 Two dice were tossed 216 times and the results were reported to be distributed as follows:

Total score: 2   3   4   5   6   7   8   9   10 11 12
Frequency: 8 20 18 20 32 30 30 27 17 11   3
Determine whether these results are likely to have come from fair
dice and fair throws.   (AEB)

6 Four treatments for a common skin disease were tested on a number
  of patients, with results as shown. Is there any evidence of a difference
  between the effectiveness of the treatments?

Treatment: A   B   C   D
Patients cured: 36   53   8   32
Total treated: 60   80   20   40   (AEB)

7 Two different processes were tried out in the manufacture of an
  article. When process A was used, 17 articles out of 120 did not
  meet specifications, while with process B, 7 out of 150 did not meet
  specifications. Draw up a 2×2 table to compare the processes, and
  discuss the difference between them.   (O & C)

8 In a sales campaign, two different brochures, A and B, were produced
  in order to advertise the same commodity. One or other of the brochures was sent to each of 500 potential buyers chosen at random.
  The number of cases in which there was a response to each brochure
  is shown in the following table:

	Response	No response	Total
Brochure A	42	158	200
Brochure B	88	212	300

Is the relative success of brochure B significant?   (O & C)

9 Two hundred volunteers were divided randomly into two equal
  groups. One group was given a new type of sleeping pill, the other a
  placebo (i.e. a dummy pill) so that no one knew which he had. After
  a suitable test, each person was asked whether he slept better, the
  same or worse, and the figures were as shown below. Are the pills
  effective?

	Better	The same	Worse
Pill:	52	32	16
Placebo:	37	35	28

10 In three firms, the numbers of employees in three categories were as
   follows:

Firm:	A	B	C
Skilled manual:	28	42	50
Unskilled:	33	76	71
Non-manual:	39	82	79

Apply a suitable test to find whether there is a significant difference between the proportions of the various kinds of workers, and state your conclusions.   (AEB)

11 The following fictitious data refers to hair and eye colours for 200 persons:

| | Hair colour | | | |
	Dark	Medium	Fair	Total
Eye colour ⎧ Blue:	3	42	30	75
Grey:	2	18	5	25
Brown:	45	40	15	100
Total:	50	100	50	200

Examine the hypothesis that there is no association between hair and eye colours.   (AEB)

12 Two hundred candidates for an oral examination go randomly to one of three examiners. The numbers of candidates in the categories credit, pass and fail are as shown in the table.

Examiner:	A	B	C	Total
Credit:	10	5	13	28
Pass:	31	38	28	97
Fail:	29	20	26	75
Total:	70	63	67	200

Examine the hypothesis that the examiners do not differ in their standards of awards.   (AEB)

13 An analysis of road accidents in a certain area was made for three successive years, and the accidents were classified as fatal, serious (if injury was involved), or minor. The figures were as follows:

	Fatal	Serious	Minor	Total
Year 1:	27	81	164	272
Year 2:	38	83	148	269
Year 3:	34	97	192	323
Total:	99	261	504	864

Examine the hypotheses:
(a) that there was no significant change in the number of fatal accidents over the period;
(b) that there was no significant change in the total number of accidents over the period;
(c) that there was no significant change in the average severity of accidents over the period.

14 Two groups of patients of different nationalities were given blood tests to find their Rhesus classifications, and the results were as shown below. Carry out a significance test and comment on the result.

Classification	First group	Second group	Total
$R_1 r$	96	165	261
$R_1 R_1$	80	99	179
rr	22	78	100
R_2	17	76	93
$R_1 R_2$	51	131	182
Others	34	51	85
Total:	300	300	600

15 In a test of car components for which the failure rate is thought to be constant, the numbers of failures in 100-kilometre intervals are shown in the following table:

Number of failures:	0	1	2	3	4	5	6	Total
Frequency:	22	27	20	19	5	2	1	96

Calculate the mean number of failures per interval, and the expected distribution of failures for a Poisson distribution with this mean. Agreement between the sets of observed and expected frequencies is taken as evidence of the constancy of the failure rate; use the χ^2-test to compare the two sets, and show that there is no significant difference between the observed and expected results.   (O & C)

Chapter 14

Miscellaneous topics

Probability density functions

If the shape of a continuous probability distribution is expressed in the form of an equation $y = f(x)$, then the probability that a random variate will be in the range x to $x + \delta x$ is equal to $y\delta x$, the area under that part of the curve. The function $f(x)$ is known as the **probability density function** (p.d.f.), since the ordinate y represents the density.

If the range of x is from h to k, the formula for the mean, usually $\bar{x} = \Sigma f x / \Sigma f$, is now given by

$$\mu = \int_h^k xy\,dx. \tag{14.1}$$

The f is replaced by $y\,dx$, and

$$\Sigma f = \int_h^k y\,dx = 1.$$

Similarly, the variance is given by

$$\sigma^2 = \int_h^k (x - \mu)^2 y\,dx. \qquad [\text{Compare } (1/n)\Sigma fd^2.]$$

It is often convenient to take the y-axis at the mean value of x, particularly if the curve is symmetrical. The equation then takes the simpler form

$$\sigma^2 = \int_h^k x^2 y\,dx.$$

An alternative form for the general case is

$$\sigma^2 = \int_h^k x^2 y\,dx - 2\mu \int_h^k xy\,dx + \mu^2 \int_h^k y\,dx$$

$$= \int_h^k x^2 y\,dx - \mu^2. \tag{14.2}$$

Again, the analogy with discrete distributions holds good.

Example 1 Find the mean and SD of the rectangular distribution represented by the density function $y = 1/k$ in the range $0 \leqslant x \leqslant k$.
The mean is

$$\int_0^k (1/k)x\,dx = \tfrac{1}{2}k,$$

which is obvious without calculation. The variance is

$$\sigma^2 = \int_0^k (1/k)x^2 dx - (\tfrac{1}{2}k)^2 = \tfrac{1}{12}k^2.$$

Hence $\sigma = k/2\sqrt{3}$ If the y-axis is transferred to $x = \tfrac{1}{2}k$,

$$\sigma^2 = \int_{-k/2}^{k/2} x^2 dx = \tfrac{1}{12}k^2 \text{ as before.}$$

Example 2 Find the SD of the parabolic distribution represented by the density function $y = 6(\tfrac{1}{4} - x^2)$ from $x = -\tfrac{1}{2}$ to $x = \tfrac{1}{2}$.
Since the distribution is symmetrical about the y-axis, $\mu = 0$. So

$$\sigma^2 = \int_{-1/2}^{1/2} x^2 \times 6(\tfrac{1}{4} - x^2)dx = 1/20, \quad \text{and} \quad \sigma = 1/(2\sqrt{5}) = 0 \cdot 224.$$

Sheppard's Correction

Example 2 above can be used to illustrate the approximation involved in grouped frequency calculations for a variance. The variance of the parabolic distribution can be estimated by dividing the area into a suitable number of sections, say 5, and using the grouped frequency method on the resulting figures. In this case the areas can be found by integration as $\frac{13}{125}$, $\frac{31}{125}$, $\frac{37}{125}$, $\frac{31}{125}$ and $\frac{13}{125}$, so that

$$\sigma^2 = \tfrac{1}{125}\{13(\tfrac{2}{5})^2 + 31(\tfrac{1}{5})^2 + 37(0)^2 + 31(\tfrac{1}{5})^2 + 13(\tfrac{2}{5})^2\}$$

$$= \tfrac{166}{3125} = 0 \cdot 05312.$$

This is a little larger than the correct value of $0 \cdot 05$.
Sheppard's Correction is a way of correcting this error, which is not an unbiased one. It is based only on the assumption that the distribution tails off smoothly to zero, and gives reasonably satisfactory results even

if it does not, as this example shows. The procedure is as follows: from the calculated variance, subtract $c^2/12$, where c is the class interval. In this case $c = 0\cdot2$, and so the correction is $0\cdot00333$. A better estimate of the variance is therefore $0\cdot05312 - 0\cdot00333 = 0\cdot04979$. The error is now about 7% of what it was before.

The correction is made necessary by two factors, which in part cancel each other out. The region of the distribution from $0\cdot1$ to $0\cdot3$, for example, is all considered to have a value $0\cdot2$; there are, in fact, more at the $0\cdot1$ end, and so the sum of squares will be overestimated. But part of this is offset by the fact that the ones near the $0\cdot3$ end will lose more by being counted as $0\cdot2$ than the $0\cdot1$ ones will gain, because of the squaring process.

In practice, Sheppard's Correction is rarely used. If the correction factor is large enough to matter, it would usually be better to change to a smaller class interval.

Probability distributions of related variables

Imagine a fairground rifle, free to move horizontally within a $90°$ angle as shown in the left-hand part of figure 14.1, so that the bullet can strike the target anywhere between A and B. Then, if the rifle is fired in a random direction, what will be the probability distribution of the bullet hole? It will not be a rectangular distribution: a small width near the end of the target is less likely to be hit than the same width at the centre, because the allowable limits of the angle are smaller.

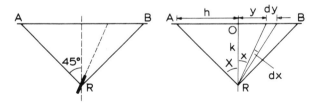

Figure 14.1   The fairground rifle problem

In mathematical form, the problem is shown in the right-hand part of figure 14.1. The variable angle x is uniformly distributed over the range $\pm X$, and the position on the target, measured by y, correspondingly varies between $\pm h$. The probability density function for x is thus $1/2X$, and the probability of an angle in the range x to $x + \delta x$ is $\delta x/2X$. If the rifle is fired within these limits of angle, the position on the target is in the range y to $y + \delta y$, and so the probability of hitting this part of the target is also $\delta x/2X$.

We require the probability distribution in a form which is in terms of y and the fixed quantities only. Since

$$x = \tan^{-1}(y/k), \qquad \delta x = k\,\delta y(k^2 + y^2)^{-1},$$

and so
$$\frac{\delta x}{2X} = \frac{k\,\delta y}{2X(k^2 + y^2)}.$$

The probability density function for the bullet hole is thus

$$F(y) = \frac{k}{2X(k^2 + y^2)},$$

which is a maximum at $y = 0$, as expected. (When $X = \pi/2$, this is known as the Cauchy distribution. It has thicker tails than the normal distribution—so much so that its SD is infinite.)

The next step is to generalise the method. Suppose we have a variable x with a known probability distribution $f(x)$ in the range $a \leqslant x \leqslant b$. (Very often this will be a rectangular distribution, so that $f(x) = (b-a)^{-1}$.) Now consider the p.d.f. of another variable y, which is functionally related to x, thus: $y = g(x)$—this corresponds to $y = k\tan x$ of the previous example. Then the p.d.f. of y, say $F(y)$, will have a range of y from $g(a)$ to $g(b)$. It is necessary that dy/dx has the same sign throughout; if this is not so, the sections with the same sign must be treated separately. Example 4 below requires this procedure.

The probability that x is in the range x to $x + \delta x$ is $f(x)\delta x$, by definition of the p.d.f.; similarly, the probability that y is in the range y to $y + \delta y$ is $F(y)|\delta y|$, the modulus sign being included because δy may be negative. Since these probabilities are the same,

$$f(x)\delta x = F(y)|\delta y|,$$

and so
$$F(y) = f(x)\left|\frac{dx}{dy}\right| = f(x)\bigg/\left|\frac{dy}{dx}\right| = \left|\frac{f(x)}{g'(x)}\right|. \qquad (14.3)$$

This can be used in the rifle problem to give directly

$$F(y) = \frac{1}{2X}\frac{d}{dy}\left(\tan^{-1}\frac{y}{k}\right) = \frac{k}{2X(k^2 + y^2)}.$$

Example 3 An equilateral triangle ABC has its altitude AD of length X. A point P is randomly chosen on AD so that AP $= x$, and a line is drawn through P parallel to BC to meet the sides at Q and R. Find the probability distribution for the area of triangle AQR, and hence find the probability that the area is less than $\frac{1}{2}\triangle$ABC.

The area y of $\triangle$AQR equals $x^2/\sqrt{3}$, and the p.d.f. for x is $1/X$. Hence the p.d.f. for y is $F(y) = \sqrt{3}/(2Xx)$.

The probability that y is in the range 0 to $\frac{1}{2}\triangle$ABC is

$$\int_0^{X^2/2\sqrt{3}} F(y)dy = \int_0^{X^2/2\sqrt{3}} \frac{\sqrt{3}dy}{2Xx}.$$

This can be worked out in terms of either x or y, and a little inspection will show that x is much easier. The dy is replaced by $(2/\sqrt{3})x\,dx$, and the limits are changed to those for x, which are 0 and $X/\sqrt{2}$. (The top limit is found from the solution of $X^2/2\sqrt{3} = x^2/\sqrt{3}$.) The integral then simplifies to

$$\int_0^{X/\sqrt{2}} \frac{1}{X}dx = 1/\sqrt{2},$$

which is clearly easier than using y.

This result can be derived directly. Since $F(y) = f(x)\left|\dfrac{dx}{dy}\right|$, $\int F(y)dy$ can be replaced by $\left|\int f(x)dx\right|$, the limits now being those of x. In this example, $f(x) = 1/X$, and so the probability that y is in the range required is

$$\int_0^{X/\sqrt{2}} \frac{1}{X}\,dx.$$

The general method is as follows. To find the probability that y is in the range y_1 to y_2, first find the values x_1 and x_2 which correspond to y_1 and y_2, and then the required probability is

$$\left|\int_{x_1}^{x_2} f(x)\,dx\right|,$$

where $f(x)$ is the probability density function for x.

Similarly, the mean of the probability distribution for y is

$$\left|\int_h^k y f(x)\,dx\right|,$$

and the variance is

$$\left|\int_h^k y^2 f(x)\,dx\right| - \left\{\int_h^k y f(x)\,dx\right\}^2.$$

The median is the solution of the equation

$$\int_{y_1}^{m} F(y)dy = \frac{1}{2},$$

which may be worked out quite easily since $|F(y)dy| = f(x)dx$.

Example 4 Through the point P on a circle radius k, a chord is drawn making angle x with the diameter through P, where x is chosen randomly in the range $\pm \pi/2$. Find the probability distribution for the length y of the chord, the mean length, and the probability that $y > k$.

The equation connecting x and y is $y = 2k \cos x$, and dy/dx is $-2k \sin x$. Thus dy/dx changes sign at the centre of the range of x, so we consider separately the cases $-\pi/2 \leqslant x \leqslant 0$ and $0 < x \leqslant \pi/2$. The p.d.f.'s are $f(x) = 1/\pi$ and $F(y)$ given by

$$F(y) = \frac{f(x)}{dy/dx} = \frac{1}{2k\pi \sin x}.$$

The mean is

$$\left| \int_{-\pi/2}^{0} y f(x)\, dx \right| + \left| \int_{0}^{\pi/2} y f(x)\, dx \right| = 2 \int_{0}^{\pi/2} (1/\pi)2k \cos x\, dx = 4k/\pi.$$

Finally, the probability that $y > k$ is

$$2 \int_{0}^{\pi/3} (1/\pi)dx = \frac{2}{3},$$

as is fairly obvious from the geometry of the diagram. The limit $x = \pi/3$ is the positive solution of $k = 2k \cos x$.

Exercises

Probability density functions

1 A continuous distribution is formed by joining by straight lines the points $(\pm 1, \frac{1}{4})$ and $(\pm 3, 0)$. Find its variance.

2 A continuous distribution is formed by joining by straight lines the points $(0, \frac{2}{3})$, $(\pm 1, \frac{1}{6})$ and $(\pm 2, 0)$. Find its variance.

3 A continuous distribution is formed by the density function $y = (3/2)(1-|x|)^2$ between $x = 1$ and $x = -1$. Find its variance.

4 An asymmetrical continuous distribution is formed by joining by straight lines the points $(-\frac{1}{2}, 0)$, $(0, 1)$ and $(1\frac{1}{2}, 0)$. Find its mean and variance.

5 (This requires integration by parts.) A continuous distribution is formed by the density function $y = \frac{1}{2}\cos x$ between $x = \pm\pi/2$. Find its variance.

6 A garage is supplied with petrol once a week. The weekly demand x, in units of 10 m³, has the continuous probability density function $f(x) = 5(1-x)^4$, if x is between 0 and 1, and $f(x) = 0$ otherwise. What must be the capacity of the petrol tank if the probability that it will be exhausted in a particular week is to be only 0·01? (JMB)

7 Given the continuous frequency density function $f(x) = 2/x^2$, where $1 \leqslant x \leqslant 2$, determine the mean and variance of x, and find the probability that x exceeds 1·5. Calculate also the median and quartile values of x, and state the interquartile range. (JMB)

8 A probability distribution for failure times is given for time t by the density function $f(t) = (1/k)e^{-t/k}$, for all positive values of t. Show that k is the mean time to failure, and that the variance is k^2. Two components in a machine have failure time distributions corresponding to means k and $2k$ respectively. The machine will stop if either component fails, and the failures of the two components are independent. Show that the probability of the machine continuing to operate for a time k from the start is $e^{-3/2}$. (O & C)

Sheppard's Correction

9 Find the variance of the triangular distribution represented by the density function $y = 1 - |x|$ for $-1 \leqslant x \leqslant 1$. Divide the distribution into five sections of equal width, and calculate the probabilities for each section. Calculate the variance of this discrete distribution. Then apply Sheppard's Correction, and compare the corrected value of the variance with that of the continuous distribution.

Probability distributions of related variables

10 A ladder 5 m long rests with one end on horizontal ground, distance x from a vertical wall. The other end rests against the wall at a height y above the ground. The value of x is uniformly distributed in the range 1·0 to 1·8 m. Prove that the median value of y is 4·8 m, and find the quartile deviation of y.

11 The average speeds of a number of vehicles each making the same journey of 60 km may be assumed to be uniformly distributed over the

interval 30 km/hr to 60 km/hr. Find the probability distribution of the time t taken to make the journey. For what proportion of vehicles will t lie between 1 hour 30 minutes and 1 hour 45 minutes? (JMB)

12 A point P is chosen at random on a fixed radius of a circle of radius k, so that its distance x from the centre has a rectangular distribution over the interval 0 to k. Find the probability density function of y, the length of the chord through P that is perpendicular to the fixed radius. Determine, to 3 significant figures, the probability that y is less than k. (JMB)

13 A point P is chosen randomly on a line AB, which is of length $2k$. Show that the mean area of the rectangle whose sides are the lengths of AP and PB is $\frac{2}{3}k^2$, and that the probability that the area will exceed $\frac{1}{2}k^2$ is $\frac{1}{2}\sqrt{2}$.

14 A sector of a circle of radius k and angle x is formed into the curved surface of a cone whose vertical height is y. If x is uniformly distributed from 0 to 2π, find the mean and the median values of y.

15 P is the mid-point of side AB of a square ABCD of side k, and Q is a point chosen at random on BC. Find the probability that the length of PQ will be less than k on each of the following assumptions: (a) that all angles made by PQ with PB are equally likely; (b) that all possible distances from B to PQ are equally likely. (JMB)

16 A and B are two points on a circle of units radius and centre O, and each has a uniform probability distribution round the circumference. Prove that the mean area of triangle AOB is $1/\pi$.

17 P and Q are two points on sides AB and AC respectively of triangle ABC, and the distribution of each is uniform along the length of the side. Prove that the mean value of the area of triangle APQ is one-quarter of that of triangle ABC.

18 A chord of a circle whose radius is k is drawn parallel to a fixed diameter in such a way that the distance x of the chord from the centre is uniformly distributed in the range $0 \leqslant x \leqslant k$. Show that the mean value of the length of the chord is $\frac{1}{2}\pi k$, and that the variance of the length is $\frac{1}{12}k^2(32-3\pi^2)$.

Further exercises

Fallacies

In each of the following, there is a fallacy in the reasoning or in the use of the data; expose and explain them.

1 A soldier exposed to gunfire should seek statistical protection in a recent shell-hole, because the odds against two shells falling on the same spot are remote.

2 The ways in which three coins can fall may be divided into two kinds: three coins all the same, or two of one of the possible results and one of the other. Take out the two coins which must be alike, and the third is then equally likely to be the same as these or different; these two cases correspond to the two kinds of result first mentioned, which are therefore also equally likely.

3 Bishops have a higher average age at death than curates. Therefore in order to increase life expectancy among the clergy, all curates ought to be appointed bishops.

4 An examination of the distributions of sentence lengths in the mathematical works of C. L. Dodgson and in the Alice books of Lewis Carroll shows a highly significant difference between the means. It therefore follows that Dodgson did not (as is commonly supposed) write the Alice books under the pen-name of Lewis Carroll.

5 More pedestrians than drivers suffer injury or death in accidents caused by alcohol. In the interests of safety, therefore, heavy drinkers should drive rather than walk home.

6 A 'martingale' is an absolutely certain way of winning at Roulette. Back red with £1; if it loses, back red again with £2, then £4, £8 and so on. When red does come up (as eventually it must), the winnings will cover all previous losses with £1 left over.

7 A gambler is to stake a sum of money on a fair coin, and £1 is to be paid if a head appears on the first throw, £2 if the first head appears on the second throw, £4 if on the third throw, and so on. The expectation for the first result is £$\frac{1}{2}$, for the second £$2 \times \frac{1}{4} = $ £$\frac{1}{2}$, for the third £$4 \times \frac{1}{8} = $ £$\frac{1}{2}$, and so on. The fair stake, which equals the expectation, is the sum of the expected returns from every possible result, and is therefore infinite. (This is an old problem, known as the St Petersburg Paradox. For a clue towards exposing it, see exercise 38 below.)

8 An investigator made 15 different and superficially unrelated measurements on 50 comparable individuals, all aged about 40—salary, height, wife's age, and so on. By comparing each list with every other, he obtained $^{15}C_2 = 105$ correlation coefficients. Of these, one was significant at the 1% level, and four others were significant at the 5% level, including two negative correlations. He therefore published these conclusions.

9 The probability of throwing a six with a fair die is 1/6; in six throws, the probability is therefore 6/6, or certainty.

10 More road accidents occur which involve vehicles travelling at between 40 and 60 km/hr than at between 140 and 160 km/hr. The slow driver is therefore a menace to safety, and should either drive faster or be banned.

11 The annual death rate per 1000 inhabitants is consistently higher in Sussex than in Surrey: it is advisable, therefore, (given the choice) to live in Surrey.

12 A coin thrown 10 times showed 7 heads. If it behaved consistently, therefore, it would show about 70 heads in 100 throws; this is significantly different from the expected value of 50, and the coin is probably biased.

13 Three bags contain, respectively, two red balls, one red ball and one white, and two white. A bag is taken at random, and one ball drawn: suppose it is white. Then the other ball in that bag is equally likely to be red or white, and so the probability that the bag originally chosen contained one ball of each colour is $\frac{1}{2}$.

14 Of the 4000 final-year students at a university, 500 were women; 450 first class degrees were awarded, 150 of them to women. There is a highly significant difference between these two proportions, and it follows that women are cleverer than men.

15 The following methods of providing random digits are suggested:
(a) The digits in the body of a table of squares in a book of four-figure tables, using a pin for random selection.
(b) The same, but using the last digits only.

(c) The last digit of the page number of a book opened at random.

(d) A pack of cards, from which the court cards have been removed, are shuffled and dealt out in order, the 10, of course, counting as zero.

16 A random sample of houses is required from a large residential area, and the following method is suggested. A large-scale map is divided into 100-metre squares, and the houses nearest to the centre of each square make up the sample, suitable allowance being made for peripheral squares.

17 Over a one-year period in a certain country, 50 million man-hours were lost through illness and 5 million through strikes. Strikes, therefore, cause less than 10% of the total absenteeism, and are thus negligible as a reason for lost production.

18 The Ruritanian census is taken for the five provinces separately, and the figures for one year were given as follows: 36750, 7366, 223600, 46720 and 89180. The total population on census day was therefore 403616.

19 The costs of a strike of car workers were calculated as follows: output of 1500 cars selling at £1000 each, cost £1500000; 2500 non-strikers laid off for four weeks, losing wages of £30 per week each, cost £300000; cut in company profits, £200000. The strike therefore cost a total of about two million pounds. (The errors here are easy to spot, but the more positive problem of estimating the true cost of a strike is much more difficult.)

20 On a certain air route, the probability that a random passenger has a bomb in his luggage is one in a million. If there are 100 passengers in an aircraft, therefore, the probability that one of them is carrying a bomb is about 1 in 10000. But the probability that two or more of them are carrying bombs is very much less—about one in fifty million, in fact. The wise traveller will therefore carry a bomb in his own luggage (unfused, of course) in order to reduce the probability that the aircraft will be destroyed by a bomb.

Probability and probability distributions

21 Three people throw a coin in turn. Show that the probabilities that each of them will throw the first head are (in order) 4/7, 2/7 and 1/7.

22 Two dice are thrown repeatedly. Find the probability that a total score of 8 will appear before a total score of 7.

23 A target consists of three concentric rings C_1, C_2 and C_3, of radii 1, 2 and 3 units respectively. A shot scores 3 points if it falls inside C_1, 2 points between C_1 and C_2, and 1 point between C_2 and C_3. If the probability density function of h, the distance of a shot from the centre, is e^{-h}, find the probability of each possible score, and show that the expected value of the score for a single shot is $3 - e^{-1} - e^{-2} - e^{-3}$. (JMB)

24 The lifetimes (x hours) of radio valves have the continuous probability density function $f(x) = 10^{-3} e^{-x/1000}$, for all positive values of x. An amplifier contains five valves of this type, and the makers guarantee that not more than two valves will have to be replaced during the first 1000 hours of use. Find the probability that the guarantee is violated, assuming that the valves wear out independently. (JMB)

25 If three points are taken at random on a circle, prove that there is a probability of $\frac{3}{4}$ that it will be possible to draw a diameter of the circle with all three points on the same side of it.

26 From a bag containing 4 red, 5 white, 6 blue and 7 black balls, four are drawn without replacement. Find the chance of their being (a) all of the same colour, (b) all of different colours. (O & C)

27 A continuous variable x is uniformly distributed in the range $0 \leqslant x \leqslant 1$. Prove that there is a probability of $\frac{1}{4}$ that two random members of the distribution will differ by at least $\frac{1}{2}$, and a probability of 9/16 that they will differ by at least $\frac{1}{4}$.

28 A lake contains a large but unknown number N of fish of a particular variety, and from the lake a random sample of m fish is taken and marked. A few days later, a second sample of n fish, which may be assumed to be random independently of the first, is taken. Find the probability that exactly k fish in the second sample are marked, and estimate the value of N which makes this probability a maximum.

29 The rounding-off errors of values in a mathematical table are rectangularly distributed between $-0 \cdot 5$ and $+0 \cdot 5$. If the table has n entries, find the probability that exactly r of them will have rounding-off errors less than x in absolute value, and give the mean and variance of r. If x is small and n is large, show that the probability that $r = 0$ is approximately e^{-2nx}. (OL)

30 In the equation $x^2 + 2x - k = 0$, k has a rectangular distribution on the interval $(0, 2)$. Find the distribution of the larger root. Find the probability that, of six equations of the above form in which the values of k are taken independently from the same rectangular distribution, no more than one has its larger root greater than $\sqrt{2} - 1$. (JMB)

31 The times t between successive breakdowns of a machine are independent, and have the probability density function Ae^{-kt} for all positive values of t. Find A in terms of k, and show that the mean of the distribution is $1/k$. Prove that the probability that the time between successive breakdowns is less than the mean does not depend on the value of the mean, and calculate this probability to 3 d.p.   (JMB)

32 A cube has side x, which is uniformly distributed in the range $1 \leqslant x \leqslant 2$ units. Find the probability density function for y, the volume of the cube, and also the mean and variance of y.

33 If an event has probability p, prove that the expectation of the number of failures before the first success is $\sum\limits_{r=0}^{\infty} rpq^r$, where $q = 1 - p$. Denoting this by E, find $E - qE$ and hence E. (This result can be extended to show that the mean number of *trials* required for a success is $1/p$, which is quite easy to prove independently.)

34 A random sample is drawn from a large population in which a proportion p have a certain rare disease. Sampling continues until a predetermined number k of the sample are found to have the disease and, at this stage, the sample size is n. Find the probability distribution of n, and show that $(k-1)/(n-1)$ is an unbiased estimate of p, i.e. that the expected value of this quantity is equal to p.   (OL)

35 A fluid contains an average of h bacteria per ml. What is the expected distribution of the number y of bacteria in samples of k ml taken from the fluid, and what conditions have to be satisfied for this distribution to arise? Each sample is put in a dish and it is observed simply whether the sample is sterile ($y = 0$) or not ($y > 0$). Write down the probability that exactly x out of n samples will be sterile, and give the mean and variance of x. For fixed n, show that this variance is a maximum when $k = (1/h)\log_e 2$.   (OL)

36 If A and B are mutually exclusive results of an experiment, with probabilities p_1 and p_2 respectively, prove that the expectation of the number of trials required to obtain at least one of both A and B is $1 + 1/p_1 + 1/p_2 - 1/(p_1 + p_2)$. (Enthusiasts are invited to apply this result to the game of Beetle.)

37 A game is played with dice as follows. The 'attacker' throws either one, two or three dice, and the 'defender' throws one. The attacker's highest score is compared with the defender's score: if it is higher, the attacker wins, and if it is equal or lower, the defender wins. Find the attacker's probability of winning in each case. (A more complicated case is also of interest: three dice against two, with highest matched with higher and second highest matched with lower; the

attacker may win two, one or none.)

38 With the data of exercise 7 above, prove that, if the resources of the bank are limited to £1 048 576 (= £2^{20}) and the rules of the gamble are amended appropriately, then the fair stake is £11.

Various topics

39 According to genetic theory, children who have one parent of blood type M and the other of blood type N will be one of three types M, MN and N, in the proportions $1:2:1$ respectively. An investigation of 200 children of such parents showed that 43 were of blood type M, 112 of type MN and 45 of type N. Discuss the agreement of these results with genetic theory.   (O & C)

40 The life in days x of an insect is such that $\log_{10} x$ is normally distributed with mean 2 and standard deviation 0·2. What is the probability that an insect will have a life of (a) more than 200 days, (b) between 50 and 150 days inclusive? Two insects have life times of t_1 and t_2, and h is the ratio t_1/t_2. What will be the distribution of $\log_{10} h$ if their life times are independent? If $h = 1\cdot8$, test whether it differs significantly from a value of 1.   (OL)

41 A hundred plots of ground in various areas were each divided into halves. To one half (chosen at random), fertiliser X was applied, and to the other half, fertiliser Y was applied. The yields from each half plot, x kg and y kg corresponding to X and Y respectively, were recorded. The following statistics were calculated.

	Mean	Variance
x	13·92	0·50
y	14·10	0·40

Assuming x and y to be independent and normally distributed, use the data to test, at the 5% level of significance, whether the fertilisers had different effects on the yield. When the value of $y-x$ was calculated for each plot, the variance of these values was found to be 0·60. Use this further information to apply an improved test to the data, and state why it is an improved test.   (JMB)

42 On each of 30 items, two measurements are made, x and y. The following summations are given: $\Sigma x = 15$, $\Sigma y = -6$, $\Sigma xy = 56$, $\Sigma x^2 = 61$, $\Sigma y^2 = 90$. Calculate the product moment correlation coefficient, and obtain the regression lines of y on x and x on y, giving the constants to 3 s.f. If the variable x is replaced by X, where $X =$

$\frac{1}{2}(x-1)$, find the correlation coefficient between X and y and the regression lines of y on X and of X on y, giving the constants to 3 s.f. (JMB)

43 Rods are manufactured to a specified length of one metre. Two batches of rods, A and B, were examined by taking a sample of six rods from each and measuring in millimetres the excess in length of each rod beyond one metre. The results were as follows:

| Batch A | 1·1 | 0·5 | 0·2 | 0·8 | 0·6 | 0·4 |
| Batch B | 0·6 | 0·9 | 1·0 | 1·2 | 1·0 | 0·7 |

(a) Compare the variances of the two batches as estimated from the samples, and comment on the result. (b) Compare the means of the two batches as estimated from the samples, and comment on the result. (c) Discuss the difference from zero of the mean of the two samples taken together. (O & C)

44 Five articles are drawn at random from a batch of manufactured articles by each of two inspectors. Each provisionally accepts the batch if he finds not more than one faulty article in his sample, and otherwise provisionally rejects it. If the two inspectors agree in their provisional verdicts, the batch is finally accepted or rejected as the case may be. If they disagree, a further sample of 10 articles is examined by the two inspectors together, and the batch is finally accepted if this further sample contains not more than one faulty article; otherwise it is rejected. If 10% of the articles in the batch are faulty, find the chance of the batch being accepted, and find also the average number of articles inspected for each batch accepted. (O & C)

45 Prove that, if the class interval in a grouped frequency distribution is less than one-third of the standard deviation, the adjustment to the SD as a result of using Sheppard's Correction is less than $\frac{1}{2}$% of the uncorrected value.

46 For a distribution of n values of x, the mean is $\bar{x}$ and the variance is s^2. If an additional value of x, denoted by X, becomes available, show that the variance of the $n+1$ values of x is given by $s'^2 = ns^2/(n+1) + nd^2/(n+1)$, where $d = X - \bar{x}$. What is the least value of s' for varying values of X? (JMB)

47 A two-stage rocket is to be fired to put a satellite into orbit. Due to variation of the specific impulse in the second stage, the velocity imparted in this stage will be normally distributed about 4095 m/s with SD 21 m/s. Find 95% confidence limits for the velocity imparted in this stage. In the first (earlier) stage, the velocity imparted will be normally distributed about 3990 m/s, with SD 20 m/s due to variation

of the specific impulse and (independently) with SD 8 m/s due to variation in the time of burning of the charge. Find 90% confidence limits for the velocity imparted in this stage. Finally, given that a final velocity of 8000 m/s is required to go into orbit and that the second stage fires immediately after the first, find the probability of achieving orbit.   (O & C)

48 At the University of Laputa, examinations are taken at the ends of the first, second and third years. At the end of the first year, 45% pass, 15% repeat the course and 40% fail and leave the university. At the end of the second year, the figures are 72%, 10% and 18%. At the end of the third year, 80% are awarded a degree, 10% repeat the third year course and 10% fail. Eight hundred degrees are awarded annually. How many new students are accepted annually, and how many students in each year are there at any time?   (AEB)

49 An examination is held in a town once each year, and is taken by all children aged between 11 and 12 on December 31st of that year. In order not to give an advantage to children born early in the year, an age allowance is given as follows: n marks are added to the total for children born in March or April, $2n$ to those in May or June, and so on, with $5n$ given to those born in November or December. Numbers of children and average marks for each group for one year were as shown in the table below. By means of a regression equation or otherwise, estimate the fairest value for n. (Work to slide rule accuracy.)

Months of birth:	J–F	M–A	M–J	J–A	S–O	N–D
Number of children:	107	89	95	110	92	107
Average mark:	167·2	168·3	156·7	151·2	145·9	146·0

50 At the end of 1970, the Ruritanian Government wanted to estimate the number of one-dollar coins in circulation. The coins had been issued on two occasions: at the end of 1950 (ten million coins) and at the end of 1960 (five million coins). It was assumed that the wastage rate due to loss, removal from the country and so on was a constant value of k% per year. A random sample of 2000 of the coins showed 1187 dated 1950 and 813 dated 1960. First, make a point estimate of the value of k, and hence give a point estimate of the number of coins in circulation at the time of sampling. Then find the 95% confidence limits for the proportions of the two issues in circulation at the time of sampling, and hence find the 95% confidence limits for the total number just found. (This is a highly simplified version of a method used in the United Kingdom in 1967, prior to the introduction of decimal currency.)

Answers to exercises

In most answers, the accuracy given is that appropriate to the data and to the nature of the calculation; often, in this kind of work, a third significant figure is of doubtful value, and sometimes even the second figure also. Greater degrees of accuracy are given in certain cases, either because the questions are of a theoretical nature or because a point of principle (such as the use of a continuity correction) would be obscured if the answer were rounded off. Where possible, a slide rule has been used for the calculations, supplemented by four-figure and occasionally five-figure tables. In a few exercises, whose numbers are marked below with an asterisk *, the answers are either not given or given only in part, almost invariably because to give the information would spoil the point of the question.

Chapter 2

1 55, 61
2 54·9
3 14 years 5 months; more exactly, 5·2 months and not 4·7
4 3·537
5 4·032
6 142·2 cm
7 7·7% increase
8 0·110 and 0·115, approximately

Chapter 3

1 20, 3
2 10, 3½
3 2·0, 0·265
4 20, 3·07
5 3, 1·49
6 4, 1·43
7 39·6, 2·33
8 0·0113, 0·00506

9 52·01, 0·476
10 −4·9, 4·48
11 5·71, 1·45
12 2·77, 0·191
13 51·2, 21·7

14 49·3 cm, 6·77 cm
15* See note on page 171
16* See note on page 171
17* See note on page 171

Chapter 4

1* See note on page 171
2 3/8, 1/4
3 5/6
4 55, 19/55
5 210, 63, 175
6 1/6, 11/36, 1/6, 5/9, 5/12
7 125/216, 75/216, 15/216, 1/216
8 10/56, 30/56, 15/56, 1/56
9 70/495, 224/495, 168/495,
   32/495, 1/495
10 1/55 for 3 or 9, 6/55 for 4 or 8,
   12/55 for 5 or 7, 17/55 for 6
11 Exact ratio is 216:1
12 1/270 725, 16/270 725,
   36/270 725
13 2/13, 1/6; 2/n, 2/(n − 1)
14 $(n + 1)/2^{n+2}$, 1/16

15 243/364
16 37/108
17* See note on page 171
18 0·253, 0·253, 0·360
19 11/216 exactly (*not* 1/18);
   73/648
20 23/35, 2/3
21* Probabilities for A, B and C
   are 36/91, 30/91 and 25/91
22 $n = 2$, 21, 47, 694, 4165, 505
   and 649 740
23 49/400 or approximately 0·12
25 0·238, 0·275
28 1/12, 1/2, 1/3, 1/12, 5/12
29 1/64, 3/32, 21/64, 9/16
30 1/4, 1/24, $1/\{4(6^{n-1})\}$, 7/10
31 $\frac{1}{2}p^3$, $\frac{1}{4}p^3$, $p^3 - \frac{3}{4}p^4$, $1\frac{1}{4}p^3 - \frac{3}{4}p^4$

Chapter 5

1 16/81, 32/81, 24/81, 8/81, 1/81
2 243/1024, 405/1024, 270/1024,
   90/1024, 15/1024, 1/1024
3 3125/7776, 763/3888
4 7/128
5 0·121
6 0·248 for 3 successes
7 0·225 for 11 or 12 successes
8 $(2n)!/\{2^{2n}(n!)^2\}$
9 0·277, 0·230, 0·128

10 37/128, 6·0
11 0·665, 0·619, 0·598, 0·637
12 5/36
13 59
14 230 (228 using four-figure
   tables)
15 0·367, 0·389
16 44
17 24
18 0·171

Chapter 6

1 0·11
2 1·0; 92, 92, 46, 15, 4, 1; 0·79, compare 1·00
3 17, 30, 27, 16, 7, 3, 1
4 0·8; 164, 131, 53, 14, 3, 0
5 0·286, 0·224, 0·132

6 0·606, 0·303, 0·076; 0·287
7* (a) 0·998, (b) 0·089
8 $K = (1 - e^{-k})^{-1}$, mean $= k(1 - e^{-k})^{-1}$
9 0·224

Chapter 7

1 0·00135, 0·0228, 0·495
2 0·0668, 0·00621, 0·595
3 0·114, 0·096
4 0·130, 0·0088
5 0·069
6 0·0004
7 0·04%, 0·35%, 51·6%
8 8700 m³/s, 10700 m³/s, 12200 m³/s

9 1·73 m
10 1·9994 cm, 0·00246 cm, 4·3%
11 0·029, 0·000005
12 1·0154 kg
13 The slow setting, by about 0·017 pence per bag
14 1·011 kg: the cost per bag is about 15·193 pence

Chapter 8

4 0·045, 0·026
5 0·0013
6 0·106, 1240 units

7 11.01 a.m., 0·92
8 0·005
9 1·9%, 4·0%

Chapter 9

1 0·397 ± 0·022
2 0·517 ± 0·015
3 0·215 ± 0·026
4 (52 ± 0·50)%
5 3·06 to 3·15
6 9700; 9600 if no continuity allowance made
7 Not significant up to $n = 120$; then significant at levels 2·2%, 4·0%, 0·4%, 0·2% and 0·01% respectively

8 Yes: significance level is less than 0·1% for one tail
9 No: significance level is about 6·5%
10 28, or more exactly at least 27·1, for a one-tail test
11 Very strong: significance level less than 0·1%
12 Probably not: significance level 0·6%
13 At least 452

14 57.4 ± 0.90
15 38.6 ± 0.66
16 71.2 ± 0.54
17 39.9 ± 0.79
18 0.5002 ± 0.00005 cm, 4%, 7%
19 108
20 96
21 664
22 Mean 30 cm, SD 2 mm; not significant, with a level of 7%
23 0.0036 cm, 5.008 ± 0.007 cm; yes, no (actually about 2.6%)
24 Not significant, but level of 9.5% suggests that further tests might prove significant
25 Evidence almost conclusive: $z = 4.2$

26 Significant at the 5% (or more exactly 3.7%) level
27 No: difference is highly significant, with a level of 0.06%
28 Difference is not significant, at a level of about 9%
29 Yes: significance level 5%
30 24.2 ± 0.4, 24.2 ± 0.6
31 13.2 ± 0.45 g
32 21.09 ± 0.73 cm, 23.8 cm
33 Not significant, even at the 10% level
34 No
35 Yes: significant at the 2% level
36 723.77 ± 0.23 units

Chapter 10

1 No: significance level 8%
2 Yes: significance level 0.2%
3 Yes, very much so: difference is over 5 SE's
4 Yes, to a surprising extent: difference is over 6 SE's
5 No: difference is less than 1 SE
6 No: significance level 11%
7 Yes, at a level of 5%
8 Yes, at a level of 0.7%
9 No: significance level about 16%
10 Yes, at a level of 2%
11 No: significance level about 20%
12 Significant difference between variances; also significant difference between means, at a level of 0.1%
13 The significance level is 0.1%
14 Difference between variances significant at the 5% level; no conclusion on the difference between the means
15 (a) Difference is not significant; (b) difference is not significant, with a level of 10%; (c) discrepancy is significant at the 5% level
16 No significant difference either between variances or between means

Chapter 11

1 0.66
2 -0.89

3 -0.45
4 $r = 0.82$; $R = 0.81$; $\tau = 0.56$

5 $r = -0.01$; $R = -0.01$;
  $\tau = 0$
6 0.57

7 -0.46, which is significant at
  the 2% level
8 $r = 0.727, R = 0.730$

Chapter 12

1 $y = 150.9 - 1.645x$; 64
2 $y = 1.18x + 20.7$; 54
3* $y = 13\frac{1}{2} - \frac{1}{8}x$, $x = 25\frac{1}{2} - \frac{1}{2}y$
4 $y = 0.0204t + 17.828$; 20.34 cm
5 $y = 18.75 - 0.796x$; $x = 18.64 - 0.821y$; $r = -0.81$

6 $y = 0.823x + 11.1$; 51 marks
7 $y = 64.8 - 0.610x$;
  $x = 100.1 - 1.505y$
8* See note on page 171
9* See note on page 171
10 $y = 1.42x + 24.3$; 70 ± 19

Chapter 13

1 See answers for chapter 9
2 See answers for chapter 10
3 Yes: significance level just over 0.1%
4 Not significant
5 Not significant
6 Not significant
7 Difference is significant at a level of just over 1%
8 Yes, at the 5% level
9 Result inconclusive

10 No significant difference
11 Significance level well below 0.1%: clear evidence of association
12 No significant difference
13 (a) Not significant;
   (b) significant at the 5% level;
   (c) not significant
14 Significance level well below 0.1%
15 $X^2 = 4.1$

Chapter 14

1 $1\frac{2}{3}$
2 $\frac{1}{2}$
3 $1/10$
4 Mean $= \frac{1}{3}$, variance $= 7/24$
5 $(\pi^2/4) - 2$ or about 0.467
6 6.02 m^3
7 Mean $= 1.386$,
  variance $= 0.0782$,
  probability $= \frac{1}{3}$;
  median $1\frac{1}{3}$, quartiles $1\frac{1}{7}$, $1\frac{3}{5}$;

  IQR $16/35$
9* Exact variance $1/6$; uncorrected variance $112/625$; corrected variance $311/1875$
10 0.058 m
11 $2/t^2$; $4/21$
12 $y/\{2k\sqrt{(4k^2 - y^2)}\}$; 0.134
14 $\pi k/4$, $\sqrt{3}k/2$
15 $0.946, 0.968$

Further exercises

22* See note on page 171
23* See note on page 171
24 0·74
26 8/1045, 24/209
28* $N = mn/k$ approximately
29* See note on page 171
30 $f(a) = a+1$, where a is the larger root; 7/64. ($p = \frac{1}{2}$ for each.)
31 $A = k$; 0·632
32 $\frac{1}{3}y^{-2/3}$, $3\frac{3}{4}$, $4\frac{9}{112}$
33 q/p or $(1/p)-1$
34 Distribution is the coefficients of t^n in $^{r-1}C_{k-1}\,p^k\,(1-p)^{r-k}t^r$, for $n \geqslant k$.
35* See note on page 171
37 5/12, 125/216, 855/1296
39 Difference is not significant
40* (a) 0·065, (b) 0·740; difference is not significant
41 Not significant, with a level of

about 6%; improved test significant, at a level of just over 2%
42* $r = 0·858$; $y = 1·102x - 0·751$; $x = 0·664y + 0·633$
43 (a) Not significant, $F = 2·08$;
(b) not significant, $t = 1·91$, with 10 d.f.;
(c) very highly significant, $t = 10$, with 11 d.f.
44 0·954; 12·1 articles per batch accepted
46* See note on page 171
47 4095 ± 41 m/s; 3990 ± 35 m/s; 0·9977
48* See note on page 171
49* See note on page 171
50 $k = 3·1\%$; estimate $8·98 \times 10^6$ coins; limits $7·9 \times 10^6$ to $10·4 \times 10^6$

Tables

Table 1.   Conversion of range to standard deviation

This table may be used to give a *rough* estimate of the population SD σ, given a sample of n observations which has range h.

n	σ/h	n	σ/h	n	σ/h	n	σ/h
2	0·886	6	0·395	10	0·325	14	0·293
3	0·591	7	0·370	11	0·315	15	0·288
4	0·486	8	0·351	12	0·307	16	0·283
5	0·430	9	0·337	13	0·300	17	0·279

Table 2.   Binomial coefficients, up to $n = 20$ and also $n = 25$

For values of x greater than 10, use $^nC_x = {^nC_{n-x}}$.

n x:	2	3	4	5	6	7	8	9	10
2	1								
3	3	1							
4	6	4	1						
5	10	10	5	1					
6	15	20	15	6	1				
7	21	35	35	21	7	1			
8	28	56	70	56	28	8	1		
9	36	84	126	126	84	36	9	1	
10	45	120	210	252	210	120	45	10	1
11	55	165	330	462	462	330	165	55	11
12	66	220	495	792	924	792	495	220	66
13	78	286	715	1 287	1 716	1 716	1 287	715	286
14	91	364	1 001	2 002	3 003	3 432	3 003	2 002	1 001
15	105	455	1 365	3 003	5 005	6 435	6 435	5 005	3 003
16	120	560	1 820	4 368	8 008	11 440	12 870	11 440	8 008
17	136	680	2 380	6 188	12 376	19 448	24 310	24 310	19 448
18	153	816	3 060	8 568	18 564	31 824	43 758	48 620	43 758
19	171	969	3 876	11 628	27 132	50 388	75 582	92 378	92 378
20	190	1 140	4 845	15 504	38 760	77 520	125 970	167 960	184 756
25	300	2 300	12 650	53 130	177 100	480 700	1 081 575	2 042 975	3 268 760

$n = 25$	$x = 11$	12
(continued)	4 457 400	5 200 300

Table 3. One-tail percentage points of the normal distribution

P	z	P	z	P	z	P	z	P	z
5·0	1·645	3·0	1·881	2·0	2·054	1·0	2·326	0·05	3·291
4·8	1·665	2·9	1·896	1·9	2·075	0·9	2·366	0·01	3·719
4·6	1·685	2·8	1·911	1·8	2·097	0·8	2·409	0·005	3·891
4·4	1·706	2·7	1·927	1·7	2·120	0·7	2·457	0·001	4·265
4·2	1·728	2·6	1·943	1·6	2·144	0·6	2·512	0·000 5	4·417
4·0	1·751	2·5	1·960	1·5	2·170	0·5	2·576	0·000 1	4·753
3·8	1·774	2·4	1·977	1·4	2·197	0·4	2·652	0·000 05	4·892
3·6	1·799	2·3	1·995	1·3	2·226	0·3	2·748		
3·4	1·825	2·2	2·014	1·2	2·257	0·2	2·878		
3·2	1·852	2·1	2·034	1·1	2·290	0·1	3·090		

A random normal variate will exceed z with probability $P/100$, and will numerically exceed z (that is, its modulus will exceed z) with probability $2P/100$. Thus P is used for a one-tail test and $2P$ for a two-tail test. This table should not be used for interpolation below $P = 1·0$.

Table 4. One-tail areas of the normal distribution

z	1·0	1·1	1·2	1·3	1·4	1·5	1·6
p	0·158 7	0·135 7	0·115 1	0·096 8	0·080 8	0·066 8	0·054 8

z	1·7	1·8	1·9	2·0	2·1	2·2	2·3
p	0·044 6	0·035 9	0·028 7	0·022 8	0·017 9	0·013 9	0·010 7

z	2·4	2·5	2·6	2·7	2·8	2·9	3·0
p	0·008 2	0·006 21	0·004 66	0·003 47	0·002 56	0·001 87	0·001 35

z	3·0	3·5	4·0	4·5	5·0	5·5	6·0
p	$1·3 \times 10^{-3}$	$2·3 \times 10^{-4}$	$3·2 \times 10^{-5}$	$3·4 \times 10^{-6}$	$2·9 \times 10^{-7}$	$2·0 \times 10^{-8}$	$1·0 \times 10^{-9}$

A random normal variate will exceed z with probability p, and will numerically exceed z with probability $2p$. This table thus gives the same information as Table 5, but in a form which is often more convenient. If interpolation is necessary, Table 5 alone should be used; after $z = 4·0$, logarithmic interpolation in this table is fairly accurate.

Table 5. The normal distribution probability function

z	0·00	0·01	0·02	0·03	0·04	0·05	0·06	0·07	0·08	0·09
0·0	0·500 0	504 0	508 0	512 0	516 0	519 9	523 9	527 9	531 9	535 9
0·1	0·539 8	543 8	547 8	551 7	555 7	559 6	563 6	567 5	571 4	575 3
0·2	0·579 3	583 2	587 1	591 0	594 8	598 7	602 6	606 4	610 3	614 1
0·3	0·617 9	621 7	625 5	629 3	633 1	636 8	640 6	644 3	648 0	651 7
0·4	0·655 4	659 1	662 8	666 4	670 0	673 6	677 2	680 8	684 4	687 9
0·5	0·691 5	695 0	698 5	701 9	705 4	708 8	712 3	715 7	719 0	722 4
0·6	0·725 7	729 1	732 4	735 7	738 9	742 2	745 4	748 6	751 7	754 9
0·7	0·758 0	761 1	764 2	767 3	770 4	773 4	776 4	779 4	782 3	785 2
0·8	0·788 1	791 0	793 9	796 7	799 5	802 3	805 1	807 8	810 6	813 3
0·9	0·815 9	818 6	821 2	823 8	826 4	828 9	831 5	834 0	836 5	838 9
1·0	0·841 3	843 8	846 1	848 5	850 8	853 1	855 4	857 7	859 9	862 1
1·1	0·864 3	866 5	868 6	870 8	872 9	874 9	877 0	879 0	881 0	883 0
1·2	0·884 9	886 9	888 8	890 7	892 5	894 4	896 2	898 0	899 7	901 5
1·3	0·903 2	904 9	906 6	908 2	909 9	911 5	913 1	914 7	916 2	917 7
1·4	0·919 2	920 7	922 2	923 6	925 1	926 5	927 9	929 2	930 6	931 9
1·5	0·933 2	934 5	935 7	937 0	938 2	939 4	940 6	941 8	942 9	944 1
1·6	0·945 2	946 3	947 4	948 4	949 5	950 5	951 5	952 5	953 5	954 5
1·7	0·955 4	956 4	957 3	958 2	959 1	959 9	960 8	961 6	962 5	963 3
1·8	0·964 1	964 9	965 6	966 4	967 1	967 8	968 6	969 3	969 9	970 6
1·9	0·971 3	971 9	972 6	973 2	973 8	974 4	975 0	975 6	976 1	976 7
2·0	0·977 2	977 8	978 3	978 8	979 3	979 8	980 3	980 8	981 2	981 7
2·1	0·982 1	982 6	983 0	983 4	983 8	984 2	984 6	985 0	985 4	985 7
2·2	0·986 1	986 4	986 8	987 1	987 5	987 8	988 1	988 4	988 7	989 0
2·3	0·989 3	989 6	989 8	990 1	990 4	990 6	990 9	991 1	991 3	991 6
2·4	0·991 8	992 0	992 2	992 5	992 7	992 9	993 1	993 2	993 4	993 6
2·5	0·993 79	993 96	994 13	994 30	994 46	994 61	994 77	994 92	995 06	995 20
2·6	0·995 34	995 47	995 60	995 73	995 85	995 98	996 09	996 21	996 32	996 43
2·7	0·996 53	996 64	996 74	996 83	996 93	997 02	997 11	997 20	997 28	997 36
2·8	0·997 44	997 52	997 60	997 67	997 74	997 81	997 88	997 95	998 01	998 07
2·9	0·998 13	998 19	998 25	998 31	998 36	998 41	998 46	998 51	998 56	998 61
3·0	0·998 65	998 69	998 74	998 78	998 82	998 86	998 89	998 93	998 97	999 00
3·1	0·999 03	999 06	999 10	999 13	999 16	999 18	999 21	999 24	999 26	999 29
3·2	0·999 31	999 34	999 36	999 38	999 40	999 42	999 44	999 46	999 48	999 50
3·3	0·999 52	999 53	999 55	999 57	999 58	999 60	999 61	999 62	999 64	999 65
3·4	0·999 66	999 68	999 69	999 70	999 71	999 72	999 73	999 74	999 75	999 76
3·5	0·999 77	999 78	999 78	999 79	999 80	999 81	999 81	999 82	999 83	999 83
3·6	0·999 84	999 85	999 85	999 86	999 86	999 87	999 87	999 88	999 88	999 89
3·7	0·999 89	999 90	999 90	999 90	999 91	999 91	999 92	999 92	999 92	999 92
3·8	0·999 93	999 93	999 93	999 94	999 94	999 94	999 94	999 95	999 95	999 95
3·9	0·999 95	999 95	999 96	999 96	999 96	999 96	999 96	999 96	999 97	999 97
4·0	0·999 97									

The figures tabulated are the areas under the normal curve from minus infinity to the given variate z—that is, the probability that a random variate will be less than z. The use of the table is explained in more detail in chapter 7.

Table 6.   Logarithms of factorials, to base 10

n	$\log n!$	n	$\log n!$	n	$\log n!$	n	$\log n!$
1	0·000 0	51	66·190 6	103	163·995 8	255	504·525 2
2	0·301 0	52	67·906 6	106	170·059 3	260	516·583 2
3	0·778 2	53	69·630 9	109	176·159 5	265	528·683 0
4	1·380 2	54	71·363 3	112	182·295 5	270	540·823 6
5	2·079 2	55	73·103 7	115	188·466 1	275	553·004 4
6	2·857 3	56	74·851 9	118	194·670 7	280	565·224 6
7	3·702 4	57	76·607 7	121	200·908 2	285	577·483 5
8	4·605 5	58	78·371 2	124	207·177 9	290	589·780 4
9	5·559 8	59	80·142 0	127	213·479 0	295	602·114 7
10	6·559 8	60	81·920 2	130	219·810 7	300	614·485 8
11	7·601 2	61	83·705 5	133	226·172 4	305	626·893 0
12	8·680 3	62	85·497 9	136	232·536 4	310	639·335 7
13	9·794 3	63	87·297 2	139	238·983 0	315	651·813 4
14	10·940 4	64	89·103 4	142	245·430 6	320	664·325 5
15	12·116 5	65	90·916 3	145	251·905 7	325	676·871 5
16	13·320 6	66	92·735 9	148	258·407 6	330	689·450 9
17	14·551 1	67	94·561 9	151	264·935 9	335	702·063 1
18	15·806 3	68	96·394 5	154	271·489 9	340	714·707 6
19	17·085 1	69	98·233 3	157	278·069 3	345	727·384 1
20	18·386 1	70	100·078 4	160	284·673 5	350	740·092 0
21	19·708 3	71	101·929 7	163	291·302 0	355	752·830 8
22	21·050 8	72	103·787 0	166	297·954 4	360	765·600 2
23	22·412 5	73	105·650 3	169	304·630 3	365	778·399 7
24	23·792 7	74	107·519 6	172	311·329 3	370	791·229 0
25	25·190 6	75	109·394 6	175	318·050 9	375	804·087 5
26	26·605 6	76	111·275 4	178	324·794 8	380	816·974 9
27	28·037 0	77	113·161 9	181	331·560 6	385	829·890 9
28	29·484 1	78	115·054 0	184	338·348 0	390	842·835 1
29	30·946 5	79	116·951 6	187	345·156 5	395	855·807 0
30	32·423 7	80	118·854 7	190	351·985 9	400	868·806 4
31	33·915 0	81	120·763 2	193	358·835 8	405	881·832 9
32	35·420 2	82	122·677 0	196	365·705 9	410	894·886 2
33	36·938 7	83	124·596 1	199	372·595 8	415	907·966 0
34	38·470 2	84	126·520 4	202	379·505 4	420	921·071 8
35	40·014 2	85	128·449 8	205	386·434 3	425	934·203 5
36	41·570 5	86	130·384 3	208	393·382 2	430	947·360 7
37	43·138 7	87	132·323 8	211	400·348 9	435	960·543 1
38	44·718 5	88	134·268 3	214	407·334 0	440	973·750 5
39	46·309 6	89	136·217 7	217	414·337 3	445	986·982 5
40	47·911 6	90	138·171 9	220	421·358 7	450	1 000·238 9
41	49·524 4	91	140·131 0	223	428· 397 7	455	1 013·519 4
42	51·147 7	92	142·094 8	226	435·454 2	460	1 026·823 7
43	52·781 1	93	144·063 2	229	442·528 0	465	1 040·151 6
44	54·424 6	94	146·036 4	232	449·618 9	470	1 053·502 8
45	56·077 8	95	148·014 1	235	456·726 5	475	1 066·877 1
46	57·740 6	96	149·996 4	238	463·850 8	480	1 080·274 2
47	59·412 7	97	151·983 1	241	470·991 4	485	1 093·694 0
48	61·093 9	98	153·974 4	244	478·148 2	490	1 107·136 0
49	62·784 1	99	155·970 0	247	485·321 0	495	1 120·600 3
50	64·483 1	100	157·970 0	250	492·509 6	500	1 134·086 4

For intermediate values, use in conjunction with ordinary log tables—e.g. $\log 102! = \log 103! - \log 103 = 161·983\ 0$.

Table 7. Two-tail percentage points of the t-distribution

$2P$: v	20%	10%	5%	2%	1%	0.1%
1	3·078	6·314	12·71	31·82	63·66	636·6
2	1·886	2·920	4·303	6·965	9·925	31·60
3	1·638	2·353	3·182	4·541	5·841	12·92
4	1·533	2·132	2·776	3·747	4·604	8·610
5	1·476	2·015	2·571	3·365	4·032	6·869
6	1·440	1·943	2·447	3·143	3·707	5·959
7	1·415	1·895	2·365	2·998	3·499	5·408
8	1·397	1·860	2·306	2·896	3·355	5·041
9	1·383	1·833	2·262	2·821	3·250	4·781
10	1·372	1·812	2·228	2·764	3·169	4·587
12	1·356	1·782	2·179	2·681	3·055	4·318
14	1·345	1·761	2·145	2·624	2·977	4·140
16	1·337	1·746	2·120	2·583	2·921	4·015
18	1·330	1·734	2·101	2·552	2·878	3·922
20	1·325	1·725	2·086	2·528	2·845	3·850
22	1·321	1·717	2·074	2·508	2·819	3·792
25	1·316	1·708	2·060	2·485	2·787	3·725
30	1·310	1·697	2·042	2·457	2·750	3·646
40	1·303	1·684	2·021	2·423	2·704	3·551
60	1·296	1·671	2·000	2·390	2·660	3·460
120	1·289	1·658	1·980	2·358	2·617	3·373
∞	1·282	1·645	1·960	2·326	2·576	3·291

The use of this table is explained in chapter 9. The probabilities are that the tabulated value will be numerically exceeded, and so refer to a two-tail test. In the case of a one-tail test, the percentages are divided by two, i.e. P rather than $2P$.

Table 7 is taken from Table III of Fisher and Yates: *Statistical Tables for Biological Agricultural and Medical Research*, published by Oliver and Boyd Ltd., Edinburgh, by permission of the authors and publishers.

Table 8. Transformation of the correlation coefficient

This is known as 'Fisher's z-transformation', and is explained in chapter 11. Note that this z is not a standardised variate.

r	z	r	z	r	z	r	z	r	z
0·10	0·100	0·72	0·908	0·90	1·472	0·972	2·127	0·991	2·700
0·20	0·203	0·74	0·950	0·91	1·528	0·974	2·165	0·992	2·759
0·30	0·310	0·76	0·996	0·92	1·589	0·976	2·205	0·993	2·826
0·40	0·424	0·78	1·045	0·93	1·658	0·978	2·249	0·994	2·903
0·45	0·485	0·80	1·099	0·94	1·738	0·980	2·298	0·995	2·994
0·50	0·549	0·82	1·157	0·95	1·832	0·982	2·351	0·996	3·106
0·55	0·618	0·84	1·221	0·955	1·886	0·984	2·410	0·997	3·250
0·60	0·693	0·86	1·293	0·96	1·946	0·986	2·477	0·998	3·453
0·65	0·775	0·88	1·376	0·965	2·014	0·988	2·555	0·999	3·800
0·70	0·867	0·89	1·422	0·97	2·092	0·990	2·647	1·000	∞

Table 9. 5% points and 1% points of the F-distribution

v_2 \ v_1	1	2	3	4	5	6	8	10	12	15	20	24	∞
1	161	199	216	225	230	234	239	242	244	246	248	249	254
	4 050	5 000	5 400	5 620	5 760	5 860	5 980	6 060	6 110	6 160	6 210	6 230	6 370
2	18·5	19·0	19·2	19·3	19·3	19·3	19·4	19·4	19·4	19·4	19·5	19·5	19·5
	98·5	99·0	99·2	99·2	99·3	99·3	99·4	99·4	99·4	99·4	99·5	99·5	99·5
3	10·1	9·55	9·28	9·12	9·01	8·94	8·85	8·79	8·74	8·70	8·66	8·64	8·53
	34·1	30·8	29·5	28·7	28·2	27·9	27·5	27·2	27·1	26·9	26·7	26·6	26·1
4	7·71	6·94	6·59	6·39	6·26	6·16	6·04	5·96	5·91	5·86	5·80	5·77	5·63
	21·2	18·0	16·7	16·0	15·5	15·2	14·8	14·6	14·4	14·2	14·0	13·9	13·5
5	6·61	5·79	5·41	5·19	5·05	4·95	4·82	4·74	4·68	4·62	4·56	4·53	4·36
	16·3	13·3	12·1	11·4	11·0	10·7	10·3	10·1	9·89	9·72	9·55	9·47	9·02
6	5·99	5·14	4·76	4·53	4·39	4·28	4·15	4·06	4·00	3·94	3·87	3·84	3·67
	13·7	10·9	9·78	9·15	8·75	8·47	8·10	7·87	7·72	7·56	7·40	7·31	6·88
8	5·32	4·46	4·07	3·84	3·69	3·58	3·44	3·35	3·28	3·22	3·15	3·12	2·93
	11·3	8·65	7·59	7·01	6·63	6·37	6·03	5·81	5·67	5·52	5·36	5·28	4·86
10	4·96	4·10	3·71	3·48	3·33	3·22	3·07	2·98	2·91	2·85	2·77	2·74	2·54
	10·0	7·56	6·55	5·99	5·64	5·39	5·06	4·85	4·71	4·56	4·41	4·33	3·91
12	4·75	3·89	3·49	3·26	3·11	3·00	2·85	2·75	2·69	2·62	2·54	2·51	2·30
	9·33	6·93	5·95	5·41	5·06	4·82	4·50	4·30	4·16	4·01	3·86	3·78	3·36
15	4·54	3·68	3·29	3·06	2·90	2·79	2·64	2·54	2·48	2·40	2·33	2·29	2·07
	8·68	6·36	5·42	4·89	4·56	4·32	4·00	3·80	3·67	3·52	3·37	3·29	2·87
20	4·35	3·49	3·10	2·87	2·71	2·60	2·45	2·35	2·28	2·20	2·12	2·08	1·84
	8·10	5·85	4·94	4·43	4·10	3·87	3·56	3·37	3·23	3·09	2·94	2·86	2·42
24	4·26	3·40	3·01	2·78	2·62	2·51	2·36	2·25	2·18	2·11	2·03	1·98	1·73
	7·82	5·61	4·72	4·22	3·90	3·67	3·36	3·17	3·03	2·89	2·74	2·66	2·21
∞	3·84	3·00	2·60	2·37	2·21	2·10	1·94	1·83	1·75	1·67	1·57	1·52	1·00
	6·63	4·61	3·78	3·32	3·02	2·80	2·51	2·32	2·18	2·04	1·88	1·79	1·00

For each pair of values given, the upper figure will be exceeded with probability 5%, and the bottom one with probability 1%. v_1 must correspond with the sample having the greater mean square deviation.

Table 10. Percentage points of the χ^2 distribution

The probabilities are that the given values will be exceeded.

v P:	99%	95%	10%	5%	2.5%	1%	0.1%
1	0.000 16	0.003 9	2.706	3.841	5.024	6.635	10.83
2	0.020 1	0.103	4.605	5.991	7.378	9.210	13.82
3	0.115	0.352	6.252	7.816	9.351	11.35	16.27
4	0.297	0.711	7.780	9.488	11.14	13.28	18.47
5	0.554	1.15	9.236	11.07	12.83	15.08	20.51
6	0.872	1.64	10.64	12.59	14.45	16.81	22.46
7	1.24	2.17	12.02	14.07	16.02	18.49	24.36
8	1.65	2.73	13.36	15.51	17.53	20.09	26.13
9	2.09	3.33	14.68	16.92	19.02	21.67	27.89
10	2.56	3.94	15.99	18.31	20.48	23.21	29.59
12	3.57	5.23	18.55	21.03	23.34	26.22	32.91
15	5.23	7.26	22.31	25.00	27.49	30.58	37.70
20	8.26	10.85	28.41	31.41	34.17	37.57	45.32
25	11.52	14.61	34.38	37.65	40.65	44.31	52.62
30	14.95	18.49	40.26	43.77	46.98	50.89	59.70
40	22.16	26.51	51.81	55.76	59.34	63.69	73.40
50	29.71	34.76	63.17	67.50	71.42	76.15	86.66
60	37.48	43.19	74.40	79.08	83.30	88.38	99.61
70	45.4	51.7	85.5	90.5	95.0	100.4	112.3
80	53.5	60.4	96.6	101.9	106.6	112.3	124.8
90	61.7	69.1	107.6	113.1	118.1	124.1	137.2
100	70.1	77.9	118.5	124.3	129.6	135.8	149.5

Part of Table 10 is taken from Table IV of Fisher and Yates: *Statistical Tables for Biological Agricultural and Medical Research*, published by Oliver and Boyd Ltd., Edinburgh, by permission of the authors and publishers.

Table 11. The negative exponential function, e^{-x}

x	0.01	0.02	0.03	0.04	0.05	0.06	0.07	0.08	0.09
e^{-x}	0.990 0	0.980 2	0.970 4	0.960 8	0.951 2	0.941 8	0.932 4	0.923 1	0.913 9

x	0.1	0.2	0.3	0.4	0.5	0.6	0.7	0.8	0.9
e^{-x}	0.904 8	0.818 7	0.740 8	0.670 3	0.606 5	0.548 8	0.496 6	0.449 3	0.406 6

x	1.0	1.1	1.2	1.3	1.4	1.5	1.6	1.7	1.8	1.9
e^{-x}	0.367 9	0.332 9	0.301 2	0.272 5	0.246 6	0.223 1	0.201 9	0.182 7	0.165 3	0.149 6

x	2.0	2.1	2.2	2.3	2.4	2.5	2.6	2.7	2.8	2.9
e^{-x}	0.135 3	0.122 5	0.110 8	0.100 3	0.090 7	0.082 1	0.074 3	0.067 2	0.060 8	0.055 0

x	3.0	4.0	5.0	6.0	7.0	8.0	9.0	10.0
e^{-x}	0.049 79	0.018 32	0.006 738	0.002 479	0.000 912	0.000 335	0.000 123	0.000 045 4

For further values, use two or three of the tabulated figures together, e.g. $e^{-3.72} = e^{-3} \times e^{-0.7} \times e^{-0.02}$.

Table 12.   Random sampling digits, 2 000 in groups of four

5 568	0 813	9 599	8 781	7 973	8 801	0 529	0 060	4 049	6 477
1 570	2 557	3 603	9 014	7 268	9 575	4 471	3 283	0 521	5 984
5 392	9 184	1 587	9 829	7 587	8 451	7 862	5 077	5 683	9 062
0 282	2 355	1 630	9 600	3 494	1 785	3 437	1 136	8 557	2 139
2 784	4 079	6 617	7 649	5 164	7 169	8 351	7 303	3 814	8 156
2 848	6 085	0 016	5 328	4 591	2 931	7 557	0 680	6 782	1 311
4 236	5 293	2 612	9 972	4 985	1 876	0 309	2 511	9 500	0 750
0 421	1 509	0 166	4 833	0 972	5 984	8 486	4 816	3 981	5 635
0 697	3 993	6 510	6 562	3 127	3 206	5 709	3 722	3 271	4 458
7 426	5 641	5 306	9 213	7 695	0 360	4 146	6 075	1 573	2 914
3 117	2 065	2 957	8 777	8 113	4 243	5 186	5 530	4 552	8 780
7 992	4 354	9 030	9 163	3 384	8 334	0 713	5 966	3 315	2 931
9 737	9 507	8 963	2 469	8 676	3 540	5 132	5 927	0 139	8 917
6 292	0 501	6 793	7 425	2 791	7 831	7 833	0 214	7 858	7 046
4 298	3 588	9 412	6 191	1 451	4 694	7 662	3 585	3 867	6 988
5 922	2 767	4 329	7 878	3 908	1 288	9 550	7 976	1 554	4 763
0 019	0 364	9 448	7 794	1 799	4 694	1 530	2 245	4 507	8 156
7 725	2 356	4 228	0 747	9 293	2 972	1 402	2 226	9 373	7 008
2 247	7 495	5 486	1 989	9 966	1 065	3 017	1 506	1 978	4 541
7 507	3 723	8 208	4 632	3 393	9 290	1 089	2 540	2 318	2 979
1 680	2 231	8 846	5 418	0 498	5 245	7 071	2 597	2 268	0 932
1 958	0 382	9 064	3 511	7 001	6 239	6 110	0 613	1 180	2 624
5 417	8 950	4 530	2 895	5 605	9 740	6 827	6 130	8 353	3 203
1 736	3 480	4 995	8 567	7 469	1 505	3 499	3 957	2 237	7 623
7 630	2 759	0 496	0 472	8 906	6 029	3 547	0 198	1 369	3 492
5 420	3 648	8 794	3 712	1 533	3 950	4 341	5 840	4 200	3 840
7 920	9 186	6 596	1 243	3 426	6 852	5 140	5 044	2 807	7 905
0 873	6 060	1 720	0 047	3 582	0 639	6 730	7 649	7 505	2 484
0 899	3 037	5 554	2 229	9 341	6 729	5 380	2 443	3 650	7 521
3 629	4 605	8 125	6 880	6 463	5 349	7 979	4 525	5 958	9 010
5 776	6 693	8 826	7 365	6 796	9 538	3 607	4 958	9 228	4 496
8 363	9 951	6 082	8 386	1 361	1 345	1 786	3 480	2 526	8 888
5 131	3 511	5 303	3 977	7 015	0 639	8 724	1 664	1 749	6 323
6 130	3 772	6 151	7 323	8 466	2 810	4 808	8 151	8 244	5 803
0 552	7 259	8 814	1 294	8 640	5 763	0 403	6 517	3 401	3 131
1 827	4 419	5 444	9 862	9 169	8 149	5 232	2 811	3 524	6 796
4 255	3 699	5 949	8 694	6 284	2 932	5 967	8 239	6 988	5 925
4 912	7 695	3 233	5 928	9 113	9 923	4 510	5 081	2 030	2 897
9 903	4 380	8 835	8 492	6 141	6 109	8 789	6 110	9 347	5 763
6 733	1 234	0 550	6 904	6 829	7 194	9 089	4 576	2 639	9 556
4 634	5 138	9 297	4 019	9 905	1 486	5 644	9 900	7 318	8 376
1 154	4 672	9 937	9 724	9 020	9 209	9 144	2 574	0 763	1 074
4 321	2 702	6 266	8 508	3 678	4 817	1 664	5 014	1 539	6 310
3 687	2 480	9 051	5 516	6 446	3 904	9 979	6 549	2 258	2 987
1 827	8 489	1 402	5 511	7 859	1 097	3 260	8 394	5 873	5 509
5 003	3 713	8 085	4 735	7 616	7 911	2 649	9 331	6 648	9 888
2 046	9 677	5 117	3 481	3 926	0 159	1 488	1 652	7 346	0 854
6 170	7 216	1 129	7 997	5 434	9 772	6 173	0 018	5 055	6 680
8 572	9 212	3 880	6 091	6 335	4 328	5 081	4 996	6 837	7 049
7 232	7 020	0 186	1 939	5 152	4 873	0 592	5 851	1 606	1 905

Index